D0192991

STAR TREK®

I'M WORKING ON THAT

Other books by William Shatner

TekWar
TekLords
TekLab
Tek Vengeance
Tek Secret
Tek Money
Tek Kill
Man o' War
The Law of War

Believe
with Michael Tobias

Star Trek Memories
with Chris Kreski

Star Trek Movie Memories
with Chris Kreski

Get a Life!
with Chris Kreski

Star Trek: Odyssey
with Judith & Garfield Reeves-Stevens
The Ashes of Eden
The Return
Avenger

Star Trek: The Mirror Universe Saga
with Judith & Garfield Reeves-Stevens
Spectre
Dark Victory
Preserver

Coming October 2002
Star Trek: Captain's Peril

STAR TREK®

I'M WORKING ON THAT

A TREK FROM SCIENCE FICTION TO SCIENCE FACT

William Shatner

with Chip Walter

POCKET BOOKS

New York London Toronto Sydney Singapore

POCKET BOOKS, a division of Simon & Schuster, Inc.
1230 Avenue of the Americas, New York, NY 10020

ISBN: 0-671-04737-X

First Pocket Books hardcover printing August 2002

10 9 8 7 6 5 4 3 2 1

For information regarding special discounts for bulk purchases,
please contact Simon & Schuster Special Sales at 1-800-456-6798 or
business@simonandschuster.com

Designed by Jaime Putorti

Printed in the U.S.A.

Stephen Hawking when visiting the engineering set of *Star Trek: The Next Generation* noted the warp engines and said, "I'm working on that," which gave me the idea for the title of this book. Here's to Stephen Hawking.

In addition, the phrase also can be used in acknowledgment of the ongoing journey that all relationships take. And so to my dear children and my fatherly relationship and to my darling wife, Elizabeth, and my husbandly relationship . . ."I'm working on that."

To boldly go where no man has gone before.

—Star Trek (1966)

Fortune favors the bold.

—Virgil

Everything should be made as simple as possible,
but no simpler.

—Albert Einstein

Prediction is difficult, especially the future.

—Niels Bohr

PROLOGUE

THE WONDERMENT DISEASE

I wonder about things. Don't ask me why (sometimes I wonder about that too). It bugs me not to understand how things work. A big word in my life is *why?*

From what I can see, curiosity, like sex and hunger, is a congenital affliction, something we picked up long ago to ensure our survival. It's what led us wide-eyed out of the jungle and transformed our little tribe of tattered hunter-gatherers into the six billion of us who now thrash around every nook and cranny of the planet, doing everything from closing deals on Madison Avenue to sweeping streets in Beijing.

This desire to know what's around the next bend has caused us humans to do some strange things, like explore space and build earth-orbiting stations; construct machines that talk and walk or smash atoms or unravel and decode the long string of DNA molecules that makes each of us possible. The need to explore was even at the heart of a certain captain and crew who embarked, in a past version of the future, on a five-year mission "to explore strange, new worlds. . . ."

None of these ventures makes much sense on the surface;

some might even say they're crazy. But we keep at it, constantly asking questions, tinkering and jabbing, like kids playing in the mud or walking through the woods. One of the laments of the twenty-first century has been that we are answering *too* many questions, filling in too many of the blanks, rubbing all of the frontiers down to little nubs. But I don't see it that way. Mystery, from where I sit, only seems to be proliferating—like tribbles.

Where is all of this mystery coming from? In my infinite ignorance I can safely tell you I don't have a clue. But I *have* noticed something interesting: a lot of the new mysteries that are unfolding around us seem to be the result of the technologies that we continually create; innovations which are themselves—curiously—the result of curiosity. Computers, cell phones, space stations, medical breakthroughs, newfangled ways to amuse ourselves or heal ourselves or make ourselves more productive. We seem enmeshed in high technology.

I suppose we take this for granted these days, but it hasn't always been this way. A generation ago if every computer on the planet had suddenly and simultaneously self-destructed, it would have been inconvenient, but far from the end of the world. (Well, the IRS might have been upset.) Today if there was a global computer meltdown, civilization as we know it would utterly discombobulate. (If you doubt the depth of our reliance on computers, consider the $100 billion spent in the United States to deal with the Y2K problem. That's a lot of greenbacks.) Governments, corporations, stock markets, airlines would all be brought to their knees. Online pornography would dissipate in a flash! Countless *Star Trek* World Wide Web sites would warp out of existence! Communications around the world would scream to a halt. Hollywood lawyers and agents would find themselves on the freeway, suddenly cut off and horrified, clutching dead cell phones in one hand while veering wildly in their BMWs off freeways all over Southern California.

Not a pretty sight.

The point is, technologies are important. And more to the point, at least from my little corner of the planet, how we use them is extremely important. That's what got me thinking. When the very first ideas that would later become this book began percolating, I couldn't help but notice a certain familiarity in the technological leaps I was witnessing all around me. It was a kind of déjà vu. And as with all déjà vus, I couldn't put my finger on it. But after some shower-stall rumination, it hit me: Everyday the world was becoming more and more like . . . *Star Trek!* Not that I saw people walking around on the decks of starships and not that I had encountered many Klingons lately—except on the floor of *Star Trek* conventions—but I sensed that there was this decided *Trek*ian trend rippling through the world as we prepared to turn the corner on the twenty-first century.

And that set me to yet *more* wondering.

Think, for a moment, how much has changed since 1966 when Captain Kirk first uttered those fateful words, "These are the voyages of the *Starship Enterprise*." Three and a half decades ago (my how the time does fly), the average computer (in the real world) was the size of a room, and had less processing power than the late model PC currently sitting on millions of our desks (each, by the way, about to be rendered rapidly obsolete by the next model). The Internet, speech recognition software, silicon chips, and brain implants were unknown. So were pacemakers and cloning. We still dialed phones and certainly owned nothing like personal digital assistants. The most state-of-the-art telephonic device was a clunky, ill-devised thing called a Princess phone. There was no such thing as gene therapy, no virtual reality, no working robots. Yet, today, this stuff, all of it, is *common*.

Millions of us stay in touch using small, wireless gadgets that we call cell phones that bear a suspicious resemblance to *Star*

Trek communicators. And remember when the *Enterprise* was in a jam in some far-flung quadrant of the galaxy? We would consult the ship's computer for information on alien cultures, history, or science. Now an estimated 520 million people have daily access to a huge databank of information that they *also* consult regularly about anything and everything. It's called the World Wide Web.

Computers, in fact, are advancing so rapidly that some scientists believe that within a generation your average desktop model will have the same mental capacity that you do! (But will it be able to channel surf and eat Doritos from a bag at the same time?) As the human genome project advances it will not be long before noninvasive gene therapies may erase whole classes of disease, including the ultimate disease: getting old. We may yet find a way to cling to that youthful glow that Mudd's women managed to corral. And very soon we will likely enter three-dimensional virtual realities that mimic the fidelity of the real world. They won't be made of solid objects, however, but of a blizzard of zeroes and ones. Soon to follow: the emergence of androids and cyborgs (cybernetic organisms) that will combine the best (it *will* be the best, won't it) of human and silicon-based intelligence.

All of these advancements are with us today, or soon will be, and here's the interesting thing: they were all foreshadowed in *Star Trek*.

As Spock might say, "Fascinating."

My reaction? "Yikes. What the hell is going on?"

When I hooked up with my coauthor, Chip Walter (also severely afflicted with "wonderment disease"), we realized that if these technologies were appearing in the real world, someone must actually be out there making them happen. Being the bright bulbs that we are, we figured out that it might be interesting to track these folks down—whoever they are—and learn a little about how they are going about it. Since Chip was a science journalist, he had a few clues about where to begin.

It seems there is this contingent of people all around the planet who concentrate on shaping up the future. *They're called sci-en-tists*. Scientists are in the big leagues of wonderment. They take the whole concept to an entirely new level. Not simply content with being puzzled, they actually try to *do* something about it. They devote years to asking strange and fascinating questions like, "Why does time move forward, not backward? Are we alone? Can we reverse-engineer the human mind? Why doesn't Dick Clark age?" These men and women are clever devils too. They not only get to spend all of their time asking wacky questions and cooking up creative solutions, they get paid for it!

For more than a year I sat across tables and stood in labs all across the country dumbfounded and slack-jawed in the presence of these people who by some strange mental alchemy, are transforming science fiction into science fact. I've listened to what they envision, heard how they work, discussed why they do what they do, and watched the results. I have tried on and fiddled with gadgets that I didn't think existed anywhere outside of a movie set. I've been a cyborg, entered virtual worlds, talked to computers (that actually listened!), and climbed the spiral ladder of human DNA. I've stumbled around the cauldrons of high technology—the Xerox Palo Alto Research Center (PARC), Carnegie Mellon University, the MIT (Massachusetts Institute of Technology) Media Laboratory, to name a few—and gotten an industrial-strength peek into the future. It was exhilarating, mind-bending, and more than once just plain overwhelming. Some days my head had been filled with so many new ideas that I felt as though I were bleeding from the ears. I developed a new respect for science.

Prior to playing a cameo role on *Star Trek: The Next Generation*, Dr. Stephen Hawking—pictured here with "holographic" versions of Albert Einstein (Jim Morton) and Isaac Newton (John Neville)—toured the engineering set of the series. When he saw the warp drive engines, he smiled and said, "I'm working on that."
(Robbie Robinson)

One particularly famous scientist, Stephen Hawking, even inspired the title of this book. Hawking is the great British physicist and author of the bestselling book, *A Brief History of Time*. He is the Lucasian Professor of Mathematics at Cambridge University, the same faculty position that Sir Isaac Newton held three hundred years ago. He is not only considered one of the great scientists of our time, but he has impressed the world with exemplary courage. In his early twenties he was struck down with amyotrophic lateral sclerosis (ALS) also known as Lou Gehrig's disease, a crippling affliction that first confined him to a wheelchair and then systematically robbed him of nearly all of his motor abilities, including speech. Today he can only communicate the insights that his extremely sharp mind conjures by using a computer that enables him to write by using several fingers on his right hand. Through a special voice synthesizer, his writing is then converted into speech. Ironically, he has become the world's most dramatic example of a cyborg, part man, part machine.

It turns out that Hawking is also a *Star Trek* fan. Several years ago he was in California and arranged to visit the set of *Star Trek: The Next Generation*. As he toured the soundstage, he passed through the *Enterprise* engineering room and paused near the warp engines. Indicating the engines, he smiled and, in his synthesized voice, said, "I'm working on that." (By the way, Hawking appeared in *The Next Generation* episode "Descent, Part I" in a holodeck poker game where he matched wits with Data, Albert Einstein, and Sir Isaac Newton—not bad company. The mind boggles at the havoc those four could wreak on house odds in Vegas.)

The moment I heard this story I knew we had our book title. The way I saw it, that was what this whole exercise should be about: finding the people who are working on the future, and then exploring what they're up to; trying to learn how these things—that we so fancifully imagined in *Star Trek*—could actually be attained.

So, as we swing into the third millennium, that's what this book is about; who, exactly, is transforming those imaginary *Star Trek* technologies into the real thing, how are they doing it, and what does it mean?

Basically I'm taking a second voyage. Thirty-five years after embarking on my first, imaginary adventure as Captain James Kirk, I am now setting out on a new, *real* voyage of exploration as just plain old Bill Shatner. You might imagine Kirk himself striking an intrepid pose on the bridge of the *Enterprise*, restating the new mission. Ah-hem . . . "To explore strange, new concepts; to seek out new technologies and gadgets that none of us comprehend a single wit. To explain them in a way even a rock would understand. To boldly (and with a light, self-deprecating sense of humor) go where readers have rarely gone before."

Cue the music.

Having completed the tour, I can tell you there were plenty of surprises. For example, it turns out that Stephen Hawking isn't the only scientist/*Star Trek* fan out there. It seems that a lot of the scientific community was inspired by the shenanigans of my fellow crew members and me over the years. I admit I am baffled by this. When I was sitting with several robotics experts at PARC (this is where the mouse and laser printer were invented), several said they had been motivated to pursue their work in artificial intelligence and robotics by *Star Trek*.

I really was amazed.

"Did it ever occur to you," I said, trying not to sound too stunned, "how strange it is that you're doing this important work because of a *television series?* I mean it was smoke and mirrors, we didn't have a clue about how you really *did* this stuff!"

They chuckled knowing scientific chuckles, and said, "Yeah, but it's the vision of the future that makes you see the possibilities. It's those stories that get you thinking, 'Wouldn't it be cool to be one of the people that makes things like that *really* possible?' "

Another time I was talking with Neil Gershenfeld, Director of the Things That Think Research Consortium at the MIT Media Lab. He told me that there was this curious connection between *Star Trek* and much of the work being done at the lab.

"Really?" I asked. "Why?"

"Well, the number of times that we've realized that we are inventing something here in the lab from *Star Trek* is spooky. One possibility I suppose is that ideas planted by watching the show are somehow subconsciously guiding the ideas we pursue. On the other hand maybe it's just like the show, we've arrived at a lot of the same conclusions; found that there's only one good way to solve many of these problems. In other words we've all come to the same answers independently. I honestly don't know which one it is, maybe both. But whatever the case, with science fact outstripping science fiction in so many ways these days, *Star Trek* holds up amazingly well when it comes to delivering insight into where technology is headed."

I heard this kind of sentiment all over the country from scientists of all stripes. NASA engineers, artificial intelligence gurus, particle physicists, biochemists. They probably all had the wonderment disease before they saw a single *Star Trek* episode or read a page of science fiction, but seeing it must have acted like a reverse inoculation, a booster shot, that made the disease worse.

Which gets me back to the reason for this book. If these *Star Trek*-ian things *are* headed our way, I, for one, want to understand them before they run me over. I suspect that a lot of you feel the same way. I mean, the future is coming. It always does. You can either be part of it, or be roadkill. I prefer the former.

However . . . there is, I must confess, a small problem. When it comes to technology, *Star Trek*-style or otherwise, I am, shall we say, challenged. More to the point, I'm in the weeds . . . witless . . . an utter ignoramus.

Well, okay, I did understand *some* things when I was working on the series. I knew, for example, that when I walked up to the doors on the bridge of the *Enterprise* on my way to the transporter room to rescue a damsel in distress, or save another civilization, or grapple with a nasty alien, that the doors would open (except when they didn't and I ended up with my face flattened like a frying pan). But I also knew that there wasn't any stupendous technology operating the doors, it was a guy with a rope looking through a peephole. (By the way, same guy and same gadgets still open the doors today.)

That pretty much represented the extent of my technical insight. I had not a clue what it meant when I sat in my captain's chair and said, "Mr. Sulu, ahead warp factor six"; or how we scrambled atoms in one place and reassembled them in another, or how Bones was supposed to be performing medical magic by waving weird gadgets (Swedish saltshakers actually) over damaged bodies. And to be honest, I didn't care. Not in those days. It was all fantasy, I figured. All I wanted to do was make certain I hit my marks and didn't botch my lines. After all, there's only so much a guy with my bandwidth can handle.

This may also explain why you should not expect this book to be exhaustive. It is not a scientific treatise. This is a flirtation, not a marriage. Men and women have spent entire lifetimes studying just one small aspect of the sciences that are the basis of many *Star Trek* technologies. To create a warp drive requires a deep and abiding knowledge of both Einsteinian and quantum physics; the development of smart machines and robots has been stumping brilliant humans for decades; the medicine practiced by McCoy requires the unraveling of biological mysteries that have won more Nobel Prizes than the Yankees have won World Series. Whole encyclopedias could be written on any one of these subjects. But they couldn't be written by me.

On the other hand, we've made a concerted effort to explore and explain the wild concepts we encountered on this odyssey in a way that hopefully lets the light bulb go off in your head from time to time; delivering those "ah hah!" moments that make life fun. (Even I had a few.)

The point is you're going to get the common man's viewpoint of cutting edge science because by necessity I enter this exploration naked as a jaybird—technologically speaking. I bring with it no scientific insight, no equivalent of Starfleet training. You have in me, a greenhorn, armed with nothing more than a bushel of bald curiosity. Maybe this sounds odd for a man who made much of his living as a starship captain and Starfleet admiral, but there it is, no use denying it.

And yet . . .

Eternal optimist that I am, I see my technological naiveté as a plus. I figure who better to host a tour of the future "for the rest of us" than I am, the consummate anti-geek? I carry with me no baggage of preconceptions, only a passel of wide-eyed ignorance. When it comes to asking the truly dumb questions we all really, secretly want to ask, I'm your man.

Under the circumstances, therefore, I prefer to see my deplorable lack of knowledge as an asset.

There's another reason, beyond my own curiosity, that I'd like to get a handle on these issues. Enlightened self-interest. The world, whether we like it or not, is wading into innovation at record speed. Things are changing so quickly, in fact, that even the acceleration is accelerating! That means that the transformations that lie ahead are coming at exponential speed. In the next ten years we will see more change than we have seen in the past twenty, and in the five years after that far more than the previous fifty! It is very likely that the wild and creative future that *Star Trek* imagined would arrive in three and four centuries will actually show up just a few decades down the road. Some are here right now!

This hurts my brain.

But it doesn't blunt my curiosity. So if you have any interest in the future, you're in the right place. Should you decide to accept this mission and read this book, you'll be required to stay sharp and keep an open mind. You'll learn things that will bend your brain at least as much as anything *Star Trek* conjured. The technologies headed our way—and the changes that they will, in turn, set in motion—are going to make even our very latest innovations look like buggy whips and crystal-set radios by comparison. I don't know that it will all come to pass precisely as some say, but then that really isn't the point, is it? Because as we all realized right from the very first episode of *Star Trek*, it's not really the destination that matters, it's the journey.

PART ONE

GETTING AROUND

Round, round, get around, I get around. Yeah.
Get around. Ooo-Ooo-Ooo-Ooo, I get around.

—*The Beach Boys, 1964*

In which our hero explores wild ways to travel faster than the speed of light, among black holes, and through time. Is beaming up possible? Instructions for building your very own time machine included (batteries, however, are not).

I look forward to the invention of faster-than-light travel. What I'm not looking forward to is the long wait in the dark once I arrive at my destination.

—Marc Beland

The scientific theory I like best is that the rings of Saturn are composed entirely of lost airline luggage.

—Mark Russell

1

FROM HERE TO NEVERWHERE

The universe is big, really *big!*

But don't take my word for it. Consider a few of these numbers. I warn you, if you actually try to get your mind around them, they'll turn your brain to tapioca.

There are 250 billion stars in the Milky Way. The Milky Way, for you nonastronomers (like me), is the galaxy we live in. Experts who know about these things have told me that if I were to ship off from one edge of it traveling 700 million miles an hour (the speed of light), it would take me 144,000 years to get to the other side! That's a lot of years. But even more astounding than the enormity of the Milky Way itself is the fact that it represents only a tiny fraction of the universe—a droplet in an ocean of Milky Ways. *There are an estimated 100 billion galaxies out there beyond our tiny planet.* If you were to count the number of stars in the cosmos—first you would be long dead before you could count even a fraction of them—but if you *could,* you would come up with a number that has twenty zeros behind it.

And there's more . . .

Even if every one of the stars above us were crammed

together cheek by jowl; if there wasn't room to slip even a teensy silicon chip between all of the heavenly bodies in all of the galaxies, the immensity of space would still be staggering. However . . . they are not crammed together. They are spread far, far apart. The emptiness between these bodies would shame even the emptiest heads of some studio executives I know. It is so empty in fact that if I were to place you in the transporter room of the *Enterprise* and set the controls to beam you to some random location in the galaxy, the chances of you arriving *anywhere at all close* to a planet or a star or any kind of solid body, would be less than one in a billion trillion trillion.

Space is spacious.

More proof. The swiftest object we humans have created is a spacecraft called *Pioneer 10*, launched from earth way back in 1972. About twelve years ago it departed the solar system, zipping along at twenty-five miles a second, a pretty stout speed. (I'm lucky if I can go twenty-five miles *an hour* on the freeways of Los Angeles). Having left our relatively crowded solar system behind, *Pioneer 10* now finds itself sailing through a vast vacancy, as solitary as a clam. Even traveling at 90,000 miles an hour, it is moving 7,500 times slower than the speed of light!

The nearest star to Earth, other than our own sun, is Proxima Centauri, combusting 4.3 light-years away. It will take *Pioneer 10* 32,000 years to get there. And this is the *closest* star! It will take 15 billion years for it to reach the next galaxy. That's a billion with a "B." To place that number in perspective, keep this in mind: 15 billion years is the current estimated age of the universe. Everything that has *ever* happened, from the big bang to your last meal, from the extinction of the dinosaurs to the rise of alien civilizations in star systems we don't even know about—*everything* has happened in those 15 billion years. And remember there are a hundred billion galaxies roughly the size of our own out there, circling, colliding, transmogrifying.

Okay. Fine, you say. I get the picture. The universe is big and things in space are far apart. This is probably why we call it "space," Bill. But we can close those distances, right, by increasing the speed?

That's what I thought, but no. Ninety thousand miles an hour might be okay if you're going from planet to planet, but when dealing in a *Star Trek* universe we're talking *interstellar* not interplanetary travel. To handle traveling between stars, we have to kick things up into a much higher gear, say the speed of light.

Okay, so let's go the speed of light. I mean let's build a big, turbocharged mother of a starship, load it with antimatter, rev it up to light speed, and plot a course for the center of the Milky Way. Be there in no time, right?

Wrong.

Be there in *30,000 years!* This is traveling at186,300 miles a second. Of course it won't feel that way to those of us onboard the ship because of something called time dilation (more on this later). We, on the starship, would only be twenty-one years older at the end of the trip, but back on Earth, assuming there *is* an Earth, things will have changed thirty millennia worth—that's enough time for all of recorded human history to have come and gone five times. Considering that almost everything I buy these days (except sweatpants) is outdated the moment I open it, I'm betting Earth will be just a *smidgen* different than it is now.

What does all of this tell us? For one thing, if you want to trek among the stars, chugging around the galaxy at the dismal speed of light is not going to cut it. Even when moving at 186,300 miles per second (at that speed you would encircle the Earth seven and a half times in a second!), we would hardly even have gotten out of the gate before *Star Trek*'s five-year mission would have been called on account of boredom. We certainly wouldn't have been encountering an alien a week.

Nope, for star trekking, we need something even faster than light speed. We need something that is, shall we say, warped.

Warped Factors

Gene Roddenberry, *Star Trek*'s creator, was a smart guy. So when he looked at the starscape in which he had chosen to set the series, he quickly understood the inherent "spacey-ness" of space. Having been a World War II pilot himself, he certainly had some sense of speed and distance. And having devoured volumes of science fiction, he knew he wasn't the first writer to confront the problem of a huge galaxy. He also knew that being constrained to the piddling speed of light simply wouldn't do given the territory his spacefarers had to cover each week.

But there is a problem with traveling faster than light, and his name is Albert Einstein. Early in the century, after much ruminating, Einstein wrote this simple, elegant equation:

$$E=mc^2$$

In addition to reflecting cosmic realities that have made everything from lasers to computers to the atomic bomb possible, this formula set the universal speed limit at 186,300 miles per second, the speed of a beam of light. Nothing, Einstein said, could travel faster, no way, no how. More precisely, he wrote in 1905, "Velocities greater than that of light . . . have no possibility of existence."*

*There have been several laboratory experiments where scientists have attempted to outrun the speed of light. Recently in a clever experiment at the NEC Research Institute in Princeton, New Jersey, three scientists conducted an experiment where a pulse of light propagated faster than itself through a cell of cesium gas. Physicists still generally agree that no object or information has yet traveled faster than light so Einstein's statement stands, so far.

You can't go up against the leading genius of the age and expect to win, so Gene did what every other self-respecting science fiction writer this century had done before him. He made something up.

He called it warp drive.

Warp drive made it possible for *Star Trek* to skirt Einstein's universal speed limit and zip around the galaxy fast enough to knock off a thoughtful (usually) and entertaining adventure a week. Imagine the problems we would have had holding to our timetable without warp drive.

Kirk: "What's our estimated time of arrival at Tycho IV, Mr. Spock?"

Spock: "Exactly twenty thousand three hundred years, six months, three weeks, four days and seven hours, Captain."

Kirk: "Very well, break out *Star Trek XLIII: Spock Jr. Meets the Son of the Nephew of Khan* and have everyone injected with enough sodium pentothal to put them out cold for a couple millennia."

So warp drive, or something like it, was an absolute necessity. At top speed, the *Starship Enterprise* could travel exactly 199,516 times 186,300 miles per second. Damned fast. But again, just to refresh your memory about the incomprehensible dimensions of the universe, even at this speedy speed (1,380,000,000,000,000 miles per hour), it would take us eighteen days to cross the celestial territory of the United Federation of Planets (10,000 light-years across), and it would *still* require ten *years* to reach the next galaxy. It says so right in the *Star Trek Encyclopedia*. This is traveling at maximum warp to the next *nearest* galaxy, never mind the remaining 99,999,999,999 other ones. (I told you this was big.) Of course it would take no time at all to get to Proxima Centauri. In fact if you left right now, you'd arrive just inside of thirteen minutes, shorter than the average urban commute.

Gene was not the first science fiction writer to conjure up faster-than-light travel. Even by the early 1960s there had been plenty of references to it in science fiction literature going all the way back to John Campbell and his 1930s pulp magazine *Astounding Stories*. In fact it's Campbell who is credited with coining the term "warp drive."

Then there was Isaac Asimov's famous Foundation series where he had his characters jaunt around the universe at faster-than-light speeds using something called "hyper drive." In fact it was the discovery of hyperspace travel that had led to the rise of Asimov's fictional Galactic Empire in the first place. Not that he went into a whole lot of detail explaining how hyper drive worked. Here's how Asimov described the experience in the opening pages of *Foundation:*

> He [Gaal] had steeled himself just a little for the Jump through hyper-space, a phenomenon one did not experience in simple interplanetary trips. The Jump remained, and would probably remain forever, the only practical method of travelling between the stars. Travel through ordinary space could proceed at no rate more rapid than that of ordinary light (a bit of scientific knowledge that belonged among the few items known since the forgotten dawn of human history), and that would have meant years of travel between even the nearest uninhabited systems. Through hyper-space, that unimaginable region that was neither space nor time, matter nor energy, something nor nothing, one could traverse the length of the Galaxy in the interval between two neighboring instants of time . . . it ended in nothing more than a trifling jar, a little internal kick which ceased an instant before he could be sure he had felt it. That was all.

Nice passage, but not exactly advanced physics.

In the 1956 sci-fi classic *Forbidden Planet*, a movie that had enormous influence on Roddenberry, the terms hyper drive and hyperspeed were used again to describe the faster-than-light travel that got the movie's impetuous crew to the "Altair system" where they then proceeded to get into all sorts of hair-raising trouble.

In the opening credits the narrator intones (over some of the weirdest music to ever accompany a movie):

> In the final decade of the twenty-first century, men and women in rocket ships landed on the moon. By 2200 A.D. they had reached the other planets of our solar system. Almost at once there followed the discovery of hyper drive through which the speed of light was first attained and later greatly surpassed. And so at last mankind began the conquest and colonization of deep space.
>
> United Planets Cruiser C-57D now more than a year out from Earth base on a special mission to the planetary system of the great main sequence star Altair.

(Note to Cyril Hume who wrote the script: You were only 140 years off on the moon-landing prediction. We'll wait and see how accurate you are on everything else.) Once again, you can't really call this, well, rocket science.

The point is that Gene, as inspired as he was by these works, knew that this sort of vague sci-fi mumbo jumbo wouldn't do for *Star Trek*. Yes, in the earliest days of the show, Roddenberry played pretty fast and loose with the whole warp drive concept, and some unabashedly sloppy terminology was tossed around. At first it was considered nothing more than this "capability" that solved some obvious dramatic problems while it moved the *Enterprise* at high speed from one place to the other throughout the galaxy. I know that in the first pilot there was talk about hyper drive and warp factors, but no technical explanations were offered. That was fine for a pilot, but once the series was given the green light, Roddenberry and his writers were forced to become a little more specific.

Why? Well, a movie or pilot is a one-shot deal. You can slip a vague generalization or two by the audience and they might be willing to buy into it, but that's not going to fly for a weekly television series. In an ongoing story you can't escape explaining how various exotic technologies work because they keep coming up. Warp drive was certainly no exception. In fact, come to think of it, it probably came up more than any other did. Pretty regularly it seemed the warp drive engines seized up or were wrecked in battle or needed "routine maintenance," and Scotty would start yapping about how if we didn't get them fixed we were going to blow a twenty-third century gasket. More than one plot was driven by a need to get a fresh supply of dilithium crystals.

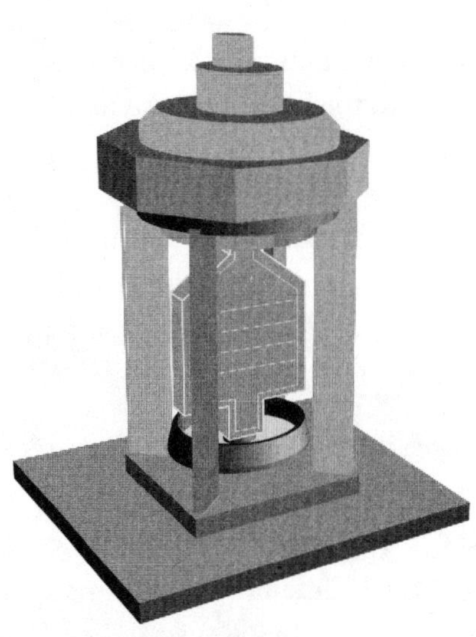

The dilithium crystal articulation frame in the *Enterprise*'s warp drive engine. In the world of *Star Trek*, the crystals control the matter/antimatter stream that enables the engines to break the laws of physics—even though Scotty kept insisting he couldn't. *(Doug Drexler)*

Naturally, being the captain of the ship, if something was wrong with the engines, I would have to ask Scotty for an explanation. That's what captains do, right? And, since the series was determined to feel real, the answer had to be plausible. I mean somehow it just wouldn't have worked if Lieutenant Commander Scott would have answered, "Well, Cap'n, the engines just keep goin' ka-chunka, ka-chunka, and if we don't fix 'em we're all going to die faster than Spock in *The Wrath of Khan*."

No, he would say something like, "Captain, you can't mix matter and antimatter cold. We'd go up in the biggest explosion . . ." That's what he told me in "The Naked Time," an episode where delusional crewman Lieutenant Kevin Riley had shut down the *Enterprise*'s engines and we suddenly found the ship overheating and spiraling in the atmosphere of the planet we were orbiting (Psi 2000).

The whole scene went something like this:

Scotty: He's turned the engines off. They're completely cold. It'll take 30 minutes to regenerate them.
Uhura: (on the intercom) The ship's outer skin is beginning to heat, Captain. Orbit plot shows we have about 8 minutes left. . . .
Kirk: . . . Captain's Log Supplemental: The *Enterprise*, spiraling down out of control. Ship's outer skin heating rapidly due to friction with planet atmosphere.
Scotty: . . . [I need] maybe 22, 23 minutes—
Kirk: Scotty, we've got six.
Scotty: Captain, you can't mix matter and antimatter cold . . .
Kirk: We can balance our engines into a controlled implosion.
Scotty: That's only a theory—it's never been done!
Kirk: Bridge. Have you found Mr. Spock yet?
Scotty: If you wanted to chance odds of 10,000 to 1, maybe assuming we had a row of computers working weeks on the right formula . . .
Uhura: Mr. Spock is not on the bridge, Captain.

Where *was* that slippery Vulcan when I needed him?

Anyhow, if you notice, this dialogue doesn't have an iota of truly technical information in it, but the overall effect of the whole conversation was that real technical issues had to be dealt with and you couldn't just snap your fingers and make them all magically disappear.

It was very effective.

Warped Thinking

But how did writers even come up with explanations like this? I asked one of the veterans of the show, D. C. (Dorothy) Fontana. Dorothy had been Gene's secretary going all the way back to his days as producer of *The Lieutenant*, an early sixties television series starring Gary Lockwood. Later she joined Gene on *Star Trek*, again as his secretary, and then, eventually, as one of the series' most valued and knowledgeable writers.

"Gene worked with several consultants," D. C. told me. "One in particular was named Harvey Lynn, a scientist with the Rand Corporation, a big think tank in Santa Monica. At first Gene didn't think so much about *how* a ship would travel from planet to planet, just that it *did*. But later he started working out the details and I'm sure the whole concept of warp drive came out of some of those discussions with Harvey Lynn. But there was never any eureka moment when Gene suddenly burst from his office and said, 'At last, I've got it! We'll call it . . . *warp drive!*' "

Lynn, it turns out, was an invaluable resource. He had been referred to Gene through Colonel Donald I. Prickett, an old Air Force buddy from his days as a pilot during World War II. "I am going to forward a copy of *Star Trek* to a physicist at Rand," Prickett wrote Gene after he had read an early summary of the series. "He's a retired AF type and I can count on him to keep it to himself—he is a creative, scientific thinker and will appreciate your concepts."

Lynn became thoroughly involved in brainstorming technical issues with Roddenberry. As Prickett had predicted, *Star Trek* was right up his alley. The Rand Corporation in Santa Monica was, and is today, a think tank that, among other things, speculates on the future. At first Lynn worked informally on the series. Later he was paid a whopping $50 per show for the use of his brain and expertise.

He contributed indispensable insights that helped shape ideas like the ship's computer (he suggested that it talk, in a woman's voice), the sickbay (he suggested that beds be outfitted with "electrical pickups" that monitor the body) and teleportation. He brought an unusual perspective to the job. He was a hardheaded scientist, but he wasn't so literal that he couldn't speculate intelligently on how you might pull something crazy off like warp drive. Gene wanted authenticity and Harvey helped deliver it.

Once Gene had realized that the *Enterprise* would need industrial strength propulsion to make its way around the galaxy, he next needed to nail down a fuel and a mechanism that infused the whole imaginary technology with that air of reality he loved so much. Nuclear power was considered, but that was ruled out. Far too puny to rev a ship up beyond light speed. Remember when the warp drive engines would go out, and Scotty would say we had to go "to impulse power." Well, impulse was another way of saying, "Turn on the nuclear engines." But nuclear power converts only a small portion of its fuel into usable energy. That's fine when dropping nuclear bombs on Earth, but not much when you want to get from here to the next star. The best those engines could do was reach a quarter of light speed.*

---

*Another form of propulsion that was being discussed in the early 1960s and that Gene may have been aware of was the Bussard Interstellar Ramjet conceived by Dr. Robert Bussard. It looks (or rather it *would* look if anyone could build it) like a huge vacuum cleaner designed to magnetically scoop up the rare, lonesome ions that drift in interstellar space, and then fuse them into nuclear fuel. The onboard reactor would convert hydrogen into helium, the same reaction that powers the sun (and every other star we know of). The genius of the machine is that it gathers its fuel as it goes, like a quarterhorse munching its way across a nice field of Kentucky bluegrass. The problem is that there's not much grass out there. Between the stars, protons are more rare than brains in a political lobby. And no matter how many of them you gather there is still no way the ship is going to exceed the speed of light. The scoop would have to be so big it would have to be made of an incredibly strong, virtually weightless (and massless) material.

In the end, when it came to moving the *Enterprise* at high speed, only one power source really made sense. Matter and antimatter. You chuckle. You say, "Sure, Bill, antimatter. Talk about mumbo jumbo!" Okay, when I first heard about it, that was my reaction too. Antimatter—right up there with flubber on the reality scale. But guess what? It *is* real; predicted in 1928 by physicist and Nobel prize–winner Paul Dirac, and first detected in 1932. It turns out that the universe is awash in antimatter.

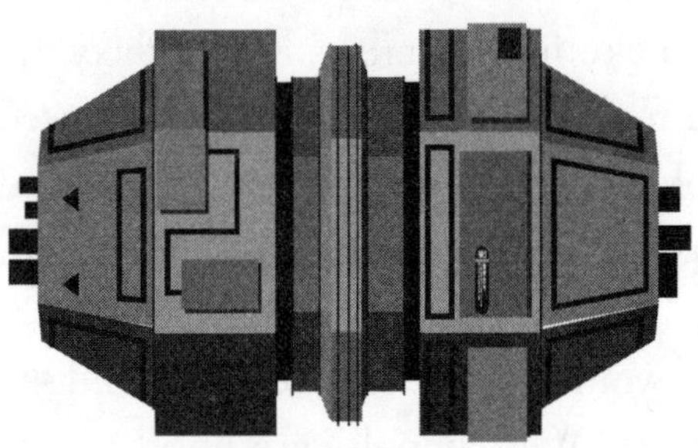

Believe it or not, antimatter does exist and is even stored away in a few very special places. This antimatter pod envisions a method for storing lots of antimatter on a Federation starship. Maybe someday we'll be able to do it for real. *(Doug Drexler)*

Every particle of matter, it seems, has an equal-partner particle of *antimatter*. The anti-particle of a negatively charged electron, for example, has a positive charge and is called a positron. Not only is antimatter real, but when you combine it with an equal amount of matter, it packs a colossal wallop, releasing every last quark of its mass as energy. *Nothing* is wasted. So handle with care because matter, *any* matter, ignites antimatter. Put enough matter in contact with antimatter and boom! there goes the planet. Compared with an antimatter reaction, a hydrogen bomb explosion is as piddling as a struck match. It is, in other words, the perfectly efficient fuel. Just what you need if you want to power up a starship built to exceed the universal speed limit.

But . . . where do you find it?

That's a problem. In the early days of the universe there were nearly equal amounts of matter and antimatter. But as it turned out, there was just a pinch more matter, not much mind you—

about one extra particle for every 100 million photons and particle/antiparticle pairs. (I looked that up.) Because matter and antimatter annihilate one another in a burst of electromagnetic radiation (energy in the form of particles called photons, to be precise) the universe we see today is dominated by the extra matter that hasn't found antimatter with which to annihilate itself.

The result: antimatter is tougher to find than an emotional Vulcan. However, I am told that physicists at the European Organization for Nuclear Research (CERN) in Switzerland have managed to manufacture antiprotons and have even stored them in supercooled magnetic fields. Just don't count on picking up a six-pack at the corner store any time soon; they can only whip up very small quantities.

But let's assume, since that's what science fiction is all about, that we can get our hands on tanks of antimatter, gobs of it. Then what? Then you need dilithium crystals and warp coils and nacelles, all of the "stuff" that makes warp engines "fly," at least in the world of *Star Trek*.

Over the years, the supporting information for all of this faux "Treknology" has become nearly as complex as you'd expect the real technology to be and has been refined to the finest of arts. After starting out as a kind of black box dramatic invention not much better defined than Asimov's hyper drive, *Star Trek*'s writers developed an increasingly detailed picture of how everything "warped" worked. Today the supporting information for these Treknologies is very impressive. A new generation of writers for ongoing *Star Trek* series can consult a fifty-one-page guide entitled the *Writers' Technical Manual* (not to be confused with the *Star Trek: The Next Generation Technical Manual*) and read highly detailed passages like this one on dilithium crystals:

> Dilithium, in its fifth-phase crystal form, is the only material yet discovered in nature [fictional *Star Trek* nature, that

> is] or manufactured which can withstand exposure to antimatter (specifically antiprotons). Its lattice structure is arranged in such a way that antimatter is held suspended in the empty spaces between the atoms when the crystal is subjected to a high-frequency electromagnetic field in the megawatt range.

The manual goes on:

> Matter and antimatter are introduced into the warp engine through separate injection reactors. . . . The crystal is placed in the path of the two streams, which would naturally collide to produce the well-known explosive reaction. Antiprotons slip through one crystal face like water through a sponge, and travel up to an opposite face. . . . The primary reaction takes place at the exit face of the crystal, at a depth of but a few atoms. Matter and antimatter undergo mutual annihilation, and the reaction is guided by the crystal. . . .
>
> Energy from the primary reaction is split into two plasma streams at equal angles from the ship's centerline. The streams are then magnetically channeled along the power transfer tubes to the warp engine nacelles.

Does any of this make sense? Could any of it work? Are you talkin' to me?

I knew after reading that passage that the time had come to seek out help. I needed to find someone who could tell me if warp speed is even possible and if so how it could be managed. Because if it isn't, to my small and mystified mind, it seemed that we humans could pretty much count on doing no more than swim around the wee pond of our own solar system until the sun blew up or *Homo sapiens* kicked off, whichever came first. At slow speed, planet trekking might be possible (we already know that), but star trekking? Not.

So I set out to find an expert, a modern-day Harvey Lynn.

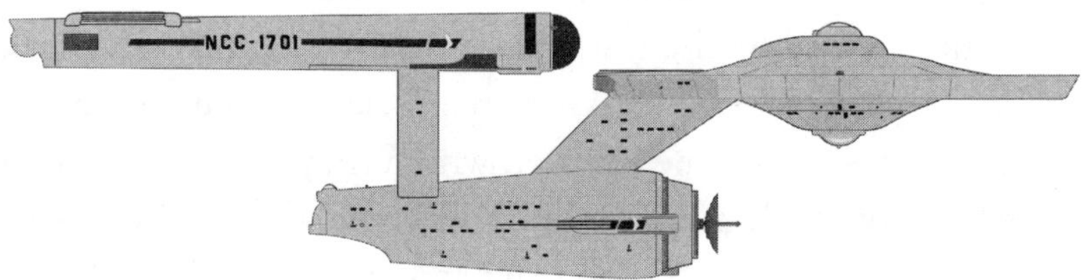

A drawing of the *U.S.S. Enterprise* NCC-1701, the one I captained way back in the future. The *Enterprise* could exceed the speed of light, but only in our dreams. However, she inspired some smart people to wonder how it might *really* be done. *(Doug Drexler)*

Warp Drive When?

It's not every day that you find a real, honest-to-God warp drive expert, but I got lucky. His name is Marc Millis. No, Marc doesn't live on the street pushing a grocery cart and he doesn't claim to have been probed by visiting aliens and he isn't walking around *Star Trek* conventions with a portable warp engine in his briefcase whispering, "Psst. Hey! Over here. Gotta blue-light special going on antimatter."

Marc is actually a *real* scientist who is *truly* investigating how to build advanced propulsion systems for NASA. In fact he is in charge of NASA's Breakthrough Propulsion Physics Program (BPP to all of you aeronautical engineers). He works out of NASA's Glenn Research Center in Cleveland, Ohio, formerly known as the Lewis Research Center. Okay, so you don't generally think of Cleveland as warp central, but it's a crazy world, and you go where the trail leads you.

Marc is a quiet, dedicated scientist who grew up fascinated with space, partly because of the Apollo program that took men to the moon and partly because of *Star Trek* and other sci-fi shows that were popular in the sixties. "I grew up with Apollo and *Star Trek* and even *Voyage to the Bottom of the Sea.* And I thought, 'Yeah, this is kind of cool. I like this.'

"But I figured by the time that I got into a career, given the progress that Apollo was making, that rockets would soon be old hat and that by the time I was grown up they would be looking for what would come after that. You know, the kind of things that were in *Star Trek.* When you're a kid, you aren't really sure how much of what you see is raw fiction and how much is based on something real. Anyhow I was very curious to figure that out."

So Millis earned his physics degree at Georgia Tech and applied at the only two NASA centers working at advanced propulsion—the Jet Propulsion Laboratory in Pasadena and the (then) Lewis Research Center in Cleveland. Lewis hired him. He went straight to work on interesting, but far from cutting edge technologies during his first several years. After all, even at NASA, any talk about something as nutty as warp drive was verboten, at least on an official level.

Unofficially, it was a different story. NASA's Glenn was loaded with science-fiction and *Star Trek* fans, and there was plenty of time passed at the cafeteria talking about any number of wild engineering ideas including warp and ion drive and wormholes. So after a while Millis started to pull together little informal groups of scientists and engineers who brainstormed wild ways to get around the galaxy in lengths of time that could be measured in years rather than millennia.

Hey, we do things all the time that were once considered impossible. Spaceships were considered nonsense a hundred years ago; now we launch satellites and shuttles and interplanetary probes every day. So, why not think about interstellar travel? It's got to come sooner or later.

So they mixed real science and engineering with some far-out science fiction and brewed up all sorts of interesting theo-

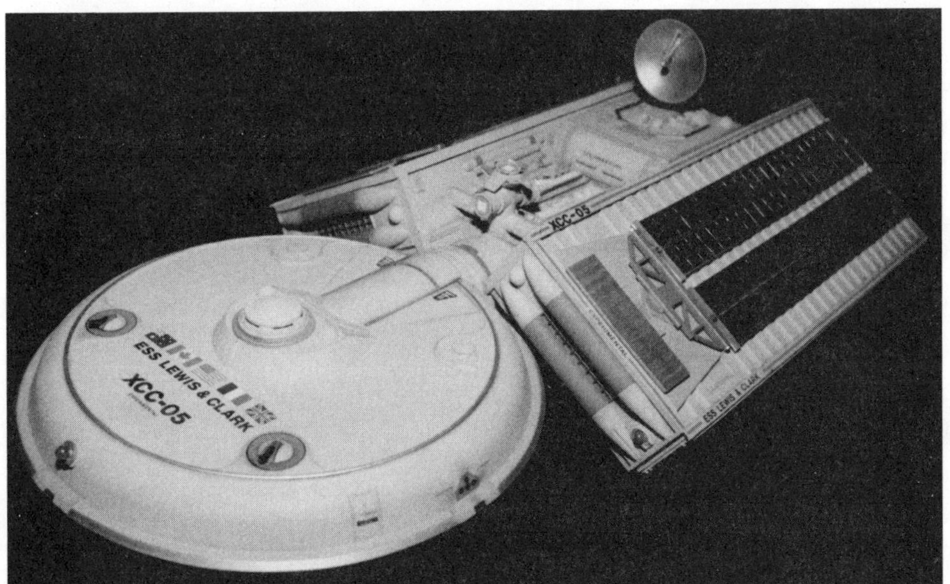

In his spare time, NASA engineer Marc Millis created a model of a ship that might someday fill the gap between the spacecraft of the twenty-first century and the *Enterprise* of the twenty-third. This ship wouldn't be capable of warp speed, but it might approach it. Millis calls her the *E.S.S. Lewis & Clark. (Marc Millis)*

ries. They even considered starting a nonprofit group called The Interstellar Propulsion Society—anything to try to advance the cause.

Help came from unexpected quarters.

In 1992 a new administrator named Daniel Goldin took over NASA. Goldin wanted to energize the space agency, give it a little boost, so to speak. He felt it had lost its sense of adventure and to capture the public imagination it needed to tackle more daring agendas. A small part of that plan was to have engineers within the agency start exploring some envelope-pushing technologies.

"Marshall Space Flight Center was asked to lead an effort to come up with long-range advanced propulsion plans," says Millis. "Their first two proposals didn't fly very well with

Goldin, who told them to be more visionary. Eventually someone from Marshall sought me out. Someone—I don't know who—had apparently asked, 'Well, what about things like faster-than-light travel and controlling gravity?' And so they tracked me down. By then I had my warp drive Web page (http://www.lerc.nasa.gov/WWW/PAO/warp.htm), and I was doing a few public talks to learn how to communicate these wild ideas in ways that would be pretty accurate and understandable.

"Anyhow, NASA proposed a really advanced propulsion program, and we decided to submit a proposal. I reformatted a lot of the work and writing our group had been doing, got a network of other people from universities to screen it, and submitted it. To make it acceptable, I tried to break it down into digestible pieces. I said, 'Okay, I'm going to prove that this is worthy one step at a time. The first step or the first question: Is there anything in the credible [scientific] literature that indicates that now is the time that we can start doing something?' And guess what? There was. It wasn't very far along, but there was plenty of material."

NASA funded Millis's proposal, and just like that (after more than a decade of preparation), he was put in charge of the BPP. Its official NASA mission: "Seek the ultimate breakthroughs in space transportation. To boldly go . . ." No, wait, wrong mission. Here we are. To tackle experiments and theories regarding the coupling of gravity and electromagnetism, the quantum vacuum, hyperfast travel, and superluminal quantum effects.

Huh?

"It comes down to this," Millis told me. "Warp drive, when?"

Well, not anytime soon. Millis doesn't pretend that he and his team are going to come up with a warp engine within the next decade or even the next few, but he figures if someone isn't out there asking these questions, how will *any* progress get made?

"It's easy to say it *can't* be done, but if we accept that, where does it get us?" asks Millis. "Literally nowhere. Progress is not made by conceding defeat.

"The way I see it, it's as important to explore these technologies as it is to attempt to figure out the age and mechanics of the universe. If we can't develop hyperfast travel, then what do we lose? We'll still have learned a lot. But if we do succeed, the benefit is huge!"

Millis says, given what we know now, he wouldn't be surprised to find it is just flat-out impossible to fly faster than the speed of light. But, he points out, when we were thinking in the 1950s about ways to go to the moon—before Apollo—there were experts that said it was impossible. They worked out their calculations and they were absolutely right—it was impossible to land on the moon in *that particular way*. You just had to change your approach, look at it from a different angle. So maybe the lesson is to redefine the problem, turn it on its head. *Then* maybe it won't look so impossible.

Big Suspenders

I decided to cut right to the chase with Marc. Why mess with pleasantries when such weighty issues were at stake? I pulled out Pocket Books's *Star Trek: The Next Generation Technical Manual* and laid out drawings of the *Enterprise*'s warp drive—dilithium crystals, warp coils, Heisenberg compensators, the whole twenty-third century shootin' match. I looked him in the eye. Did any of this make sense?

Millis took the book in hand and looked at it, and then looked up. "This?"

"Yeah, with the antimatter being combined and refined through the dilithium crystals. The (I looked at the book) magnetically channeled plasma streams, the nacelles . . . all of it."

Millis looked back at the diagrams again. If I'm not mistaken, he actually squirmed a little. "Well, the *Enterprise* is an excellent tool for inspiration," he said diplomatically, "but you definitely don't want to use it as a research guide. The motivation for putting *this* together," he said, gesturing at the manual, "is for dramatic effect. And then folks have thrown lots of things in to fill in the gaps. And they've been very clever with the terminology they use to give it more weight and greater feeling of reality."

He looked up. "But drama is not science."

"So there is some serious suspension of disbelief going on here."

"Yeah. Some big suspenders."

"Well, if you don't do it *this* way, how *do* you do it?"

Millis made one thing immediately clear: You don't outrace the speed of light by simply *propelling* the *Enterprise* down the star lanes at faster and faster speeds. What? You don't? I had always assumed that the *Enterprise*'s warp engines worked something like a supercharged space shuttle, or, simpler still, a balloon powered by escaping air that made rude noises as it flew around the room. I mean, *that* I get. Fire the engines up and they propel you like a rocket through space, breaking the light barrier the way Chuck Yeager shattered the sound barrier back in 1947.

Of course thirty-five years ago, wandering the Desilu lots where we shot *Star Trek*, I didn't wonder much about how these things might have worked. I didn't question why we weren't all splattered like paint balls around the deck of the *Enterprise* as the G-forces reached into double digits. And I certainly didn't consider the complex physics involved in all of this. All I did was sit in my captain's chair and say, "Yo, Sulu, warp factor two, and don't let the moons of Saturn hit us in the ass as we leave the solar system." So when Marc told me that traveling faster than light doesn't involve anything like the balloon propulsion concepts I had been thinking about, I couldn't get it.

He patiently explained it to me using the sound barrier as an example. When Yeager broke the sound barrier, he told me, it wasn't *sound* that broke the barrier, it was an object, the X-15 jet that he was flying. The X-15 reached speeds that were faster than the sound waves it was creating.

With light, however, it's a different story. The atoms and molecules that make up matter are connected by electromagnetic fields, *the very same stuff that light is made of*. Both are governed by the same physical laws. That means that when you try to break the speed of light, you are trying to do it with the very same forces that light itself consists of. So how can an object possibly travel faster than the force that makes it possible in the first place? Einstein said it can't. That's why he called the speed of light "constant" (the C in mc^2). Got that? Good, because I'm not sure I do. (There will be a pop quiz later.)

Dilate "T" for Time

The light-speed speed limit is one problem, but there are plenty of others. One of them is called time dilation. As a star cruiser approaches light speed, a speed so much faster than the other objects around it, time actually slows down for those on the ship. Why? Well, a few thousand people in the world might truly understand why. I'm not one of them. So I asked Millis for an answer.

He explained it like this. When one person is moving much faster than another, the whole idea of events happening simultaneously goes out the window. For example, let's say you are standing still, watching Leonard Nimoy take a bow onstage following a great performance. If I am moving at the speed of light, I will not see Leonard take his bow at the same time that you do

			Approximate Time to Travel						
SPEED	Kilometers per hour	Number of times speed of light	Earth to moon 400,000 kilometers	Across Sol system 12 million kilometers	To nearby star system 5 light-years	Across one sector 20 light-years	Across Federation 10,000 light-years	To nearby galaxy 2,000,000 light-years	NOTES
Standard orbit	9600	<0.00001 SUBLIGHT	42 hours	142 years	558,335 years	2 million years	1 billion years	223 billion years	synchronous orbit around Class-M plane
Full impulse (1/4 light speed)	270 million	0.25 SUBLIGHT	5.38 seconds	44 hours	20 years	80 years	40,000 years	8 million years	normal max impulse speed
Warp factor 1	1 billion	1	1.34 seconds	11 hours	5 years	20 years	10,000 years	2 million years	Warp 1 = SPEED OF LIGHT
Warp factor 2	11 billion	10	0.13 seconds	1 hour	6 months	3 years	992 years	198,425 years	
Warp factor 3	42 billion	39	0.03 seconds	17 minutes	2 months	1 year	257 years	51,360 years	
Warp factor 4	109 billion	102	0.01 seconds	7 minutes	18 days	2 months	98 years	19,686 years	
Warp factor 5	229 billion	214	0.006291 seconds	3 minutes	9 days	1 month	47 years	9,357 years	
Warp factor 6	421 billion	392	0.003426 seconds	2 minutes	5 days	19 days	25 years	5,096 years	
Warp factor 7	703 billion	656	0.002050 seconds	1 minute	3 days	11 days	15 years	3,048 years	
Warp factor 8	1.10 trillion	1,024	0.001313 seconds	39 seconds	2 days	7 days	10 years	1,953 years	
Warp factor 9	1.62 trillion	1,516	0.000887 seconds	26 seconds	1 day	5 days	7 years	1,319 years	
Warp factor 9.2	1.77 trillion	1,649	0.000816 seconds	24 seconds	1 day	4 days	6 years	1,213 years	normal max speed of Federation starships
Warp factor 9.6	2.05 trillion	1,909	0.000704 seconds	20 seconds	23 hours	4 days	5 years	1,048 years	
Warp factor 9.9	3.27 trillion	3,053	0.000440 seconds	13 seconds	14 hours	2 days	3 years	655 years	
Warp factor 9.99	8.48 trillion	7,912	0.000170 seconds	5 seconds	6 hours	22 hours	1 year	253 years	
Warp factor 9.9999	214 trillion	199,516	0.000007 seconds	0.2 seconds	13 minutes	53 minutes	18 days	10 years	subspace radio speed with booster relays
Warp factor 10	∞ <infinite>	∞ <infinite>	0	0	0	0	0	0	Warp 10 unattainable except with transwarp

The universe and the Milky Way are *big*. Even traveling at warp 9, it can take a long time to get around the galaxy. Roddenberry knew that without something like warp drive, *Star Trek*'s five-year mission would have been over before we even got out of the solar system. This chart illustrates just how handy warp drive is.

because I am traveling at somewhere around 186,000 miles a second. Our perception of reality is, relatively speaking, out of sync because our speeds, relative to one another, are extremely different.

This means that our perception of time is different based on how fast we are moving. In fact it means that if I am traveling faster than you are, time actually stretches out, slows down for me as compared to you. It dilates.

As ridiculous as this sounds, time dilation has actually been proven. Here's how. Let's say I have two atomic clocks (doesn't everyone?). To perform a time dilation experiment, I keep one in Los Angeles and send the other to Leonard Nimoy who happens to be in New York meeting with some high-powered publishing executives. (He's always got some deal going.) I pocket my clock and hop on a New York–bound jet. As I'm taxiing down the runway I sneak a call to Leonard and tell him to start his clock. I start mine at exactly the same time. When the jet lands in New York, Leonard greets me at the airport. After hugs and hellos and some Vulcan dancing, we compare the elapsed time on the two clocks. Low and behold, my clock, the one that has been airborne for the past five hours, has ticked off twenty-two nanoseconds less than Leonard's—exactly what Einstein himself would have predicted.

So never let anyone tell you that you can't save time by moving fast. It's a fact.

Now if you can literally wrinkle time flying on a commercial jet at a paltry five hundred miles an hour, imagine the time dilation you could rack up on the *Enterprise* which spends most of its five-year mission jamming around the galaxy at way better than the speed of light.

This could create some serious plot problems. People all over the Federation would be out of sync because almost no one would be moving through time and space at the same speed.

Birth dates would mean nothing, rendered obsolete by star travel. Those living on space stations like DS9 or on planets like Vulcan or Earth would be aging like crazy relative to those of us zipping around in our starships. You wouldn't want to make a date with someone at home and agree to meet six months later. When you showed up at the door, your date would be ancient, if not dead, and you would be young and ready to boogie. Bad combination.

Simply by doing some high-speed space traveling, we will have flung ourselves years into the future, or others equally far into the past, take your pick. As Einstein said, it's all relative.

Energy Crisis

As if we don't have enough trouble already, there remains one final problem with hyperfast travel. Energy. No matter how fast you are traveling, if you want to pick up speed, you need more energy. Simple example: you get in your car, jump onto the nearest freeway, and floor it. You're going to run out of gas a lot sooner than if you simply cruised. Why? Because the faster you go, the more fuel (energy) you need. Same with traveling among the stars. If you want to travel at the speed of light, it takes not simply more energy, or even a lot of energy, but *all of the energy there is in the universe*. This is a problem. And remember we are merely talking about light speed—a plodding 700 million miles an hour. Never mind warp-factoring multiples of light speed.

So it turns out that attaining speeds that outrun light isn't about just coming up with a real powerful fuel, lighting up the engines, and letting her rip. It's about, as Marc says, "getting traction on the surface—the very fabric—of space and time itself, like a car's tires get traction on a highway."

Accomplishing that means getting *way* beyond the universe that Newton imagined, even beyond the one Einstein imagined, certainly beyond anything I can imagine. It isn't about propulsion in the space shuttle sense. It is, if I understood what Marc told me, about flying *without* propulsion. It's about folding, spindling, and mutilating the fundamental electromagnetic forces of the cosmos—the same forces that make light possible, and hold the fabric of existence itself together. It means building a machine that can actually warp space and time around it.

But is that even possible?

Maybe.

Physics is like sex: sure, it may give some practical results, but that's not why we do it.

—Richard Feynman

I canna change the laws of physics.

—Lt. Commander Montgomery Scott

2

HONEY, I SHRUNK THE GALAXY

Every time Mexican physicist Miguel Alcubierre used to relax at his home in Wales and watch *Star Trek: The Next Generation* he was painfully aware that the *Enterprise* could not possibly be motoring around the galaxy faster than the speed of light. No matter how he racked his brain, he knew it simply couldn't be done. Einstein said so.

On the other hand, he thought it would be outrageously cool if such a thing could be done. Think of the possibilities. But everything he had ever learned told him warp drive was a no go. He didn't like that and he wished he had a solution.

Then . . .

"Then it hit me," he says. "The very name 'warp drive' means it must distort space. I knew that we already had a physical theory that told us how to distort space. It's called the Theory of General Relativity. So I sat down to 'design' a space-time distortion that would basically accomplish what science fiction calls 'warp travel.' Once I got into it, it turned out to be quite simple to come up with the required space-time geometry."

(Sounds like Spock.)

Alcubierre's warp drive theory first came to light when the editors of *Classical and Quantum Gravity* published a letter he wrote to them in 1994. He wrote it, he told me, as a kind of lark. The faster-than-light concept he described was truly creative. Not only did it eliminate the need for using all of the energy in the universe to power it, it covered enormous interstellar distances in a blink, and it disposed of all time dilation problems. Yet at no time does the ship ever even *approach* the speed of light.

Wait a minute! you say. How, if it is not traveling at better than the speed of light, can the machine possibly be traveling at warp speed? Well that's exactly what I wanted to know. Unfortunately Alcubierre is on the faculty of the Max Planck Institute for Gravitational Physics. (He was at the University of Wales at Cardiff when he wrote the paper.) I, however, live in Los Angeles. A quick face-to-face meeting just didn't seem to be in the cards. (When *are* we going to get those transporters working?) So instead I dropped him an e-mail and he graciously filled me in, digitally. I learned a lot.

Ballooniverse

Imagine the universe—all of space and time—as a huge balloon; a big black dirigible rapidly inflated at the moment of the Big Bang. Also imagine all of the stars in the universe as tiny white dots painted on this balloon. If you inflate the balloon, the dots grow farther apart. If you deflate it, they grow closer together.

Now imagine that you want to travel from one dot to the other. How do you do it? Well, you march across the surface of the balloon from Dot 1 to Dot 2 then on to Dot 3, right? Right. Except remember, these dots are stars and stars are immensely far apart so simply roaming from one to the other takes forever. Now what if, instead, you had a ship that *squeezed* the balloon,

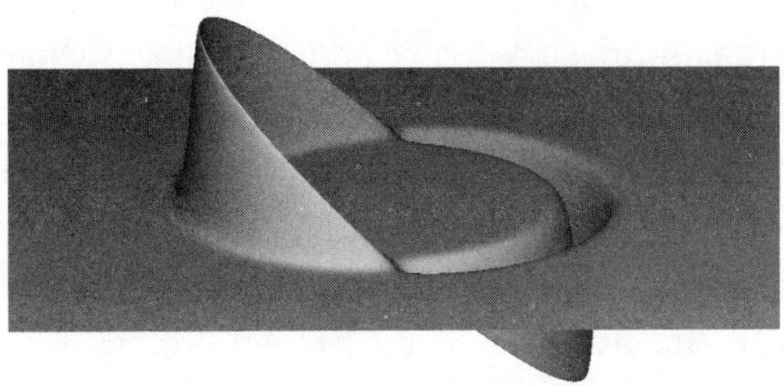

Miguel Alcubierre's "real life" warp drive engine would distort the space around it. This is a kind of cutaway image. In three-dimensional space, the ship would actually sit inside of a bubble (the curves you see would envelop it). In front of the ship, space is contracting, pulling it forward. Behind, the ship is re-expanding space so that it rides the distortion like a surfer rides an ocean wave. *(Miguel Alcubierre)*

and pulled the dots closer and closer together, until the space between the stars, became smaller. In effect, what if the ship warped space, gathering the stars toward *you* rather than propelling you to the stars? Furthermore what if, as the ship "passes" a certain location, it re-expands the balloon behind it, returning that sector of the universe to its old familiar form. (Think of it as a horizontal hourglass that gathers the sands of time and space toward the center of its neck and then re-expands them behind it.)

This is entirely different than revving a ship up to faster-than-light speed. Such a machine, in fact, would never actually exceed the speed of light because it wouldn't have to cover enormous distances, which in turn means it wouldn't violate any of the laws of general relativity. And because it doesn't exceed light speed* it would also never suffer from those pesky time dilation problems.

---

*Not only doesn't the ship exceed light speed, as far as anyone or anything in the vicinity is concerned, it doesn't appear to be moving at all!

Brilliant! Alcubierre's theoretical ship means you *can* travel across the galaxy in a reasonable period of time without trashing either Einstein's equations *or* your personal life. Go ahead, make all of the dates you want!

With all of the roadblocks that the laws of physics had thrown up; with all of the limits and hurdles I had heard about up to this point, reading about Alcubierre's warp drive was like finding out that there really is a Santa Claus. I thought, Well, this means that someday, some Captain Kirk–kind-of-guy really could sit on an Alcubierre ship, plot a course for Deep Space 9, and say, without even blinking an eye, "Ahead, warp factor three, Mr. . . . whatever your name is," and it would actually be possible. Beautiful.

But wait . . .

There is a problem. (Isn't there always?) Literally warping the universe according to your whims, I learned, takes just a *teensy, weensy* bit of energy. How much? Well . . . more power than all of the energy expended so far by the sun since it first came into existence five billion years ago. (*Now* I understand why Scotty kept saying he couldn't give me more power.)

"When you distort space," Miguel informed me in one of his e-mails, "there's a price to pay. In real life you can't just deform space any way you like. Space (and time) are shaped by the distribution of mass and energy. To get the 'warp' effect, you need an extremely strange kind of matter called 'negative energy.' "

Negative energy. I blinked.

"It's an energy field that produces gravitational repulsion instead of attraction."

What's he saying here? That warp drive is repulsive?

Okay. An analogy. Say you have a lung full of negative air. Don't ask me where you got it, let's just say you have it. Now you pick up an already inflated balloon and start blowing this negative air into it. What happens? The harder you blow, the more the balloon deflates. Negative air would do exactly the opposite of normal air.

The same, theoretically, is true with negative energy. It would do the opposite of what your garden-variety energy would do. Instead of keeping everything in place, it would let you distort Einstein's space-time continuum any way you wanted, as long as you had a hefty supply. But remember, for a starship, you need the negative energy equivalent of all of the energy expended so far by the sun since it first came into existence. What are the chances of that?

Believe it or not, energy at this titanic level is actually expended at choice locations throughout the universe. In fact this is, roughly speaking, the amount of energy that a black hole exerts every *second* of its existence. A black hole is, so far, the only thing any scientist can imagine being strong enough to squeeze space and time the way Alcubierre's starship would need to squeeze it. In other words flying the *Enterprise* would be something like flying a black hole through space (except this black hole wouldn't simply swallow space and time, it would also reinflate it after condensing it). A nifty trick if we can pull it off. Can we?

Right now, a big fat no. We have to solve that negative energy problem. But fear not, there may yet be ways to outflank the cosmic speed limit. In fact, there may even be ways to flit from one end of the galaxy to the other, let alone between the stars. There may even be ways to flit from one *time* to another or from one *universe* to another.

But you have no chance of understanding how until you first get a primer in what I (knowing nothing about any of these things) call Black Hole-ology.

Black holes are where God divided by zero.

—Stephen Wright

3

BLACK HOLE-OLOGY

If popular culture were any barometer, you'd think black holes were as common as strippers on *The Howard Stern Show*. It seems that even people that don't give a hoot about science or science fiction have heard of them. The term has been drilled into our culture; embedded in the language. You might hear someone say, "Talking to that actor is like talking to a black hole. Does any serious thought *ever* escape that mind?" Or, "When I talk to my husband, I feel like everything I say is just going into a black hole." (If he's watching a football game, operating a power tool, or you're asking him to stop for directions, I guarantee you it is.)

So what are black holes? Where did they come from? Why does everyone toss the term around like they actually know what they're talking about? Hold your horses! One thing at a time!

First, I have to fit *Star Trek* into all of this. I mean this is a *Star Trek* book.

For those of you who are not official experts on the series' history, you might be surprised to learn that *Star Trek* did nothing less than introduce millions of science fiction fans to the black holeness of the universe way back in 1967 in one of its earliest episodes.

In 1994, the Hubble Space Telescope photographed a disk of hot gas swirling at the center of a giant, elliptical galaxy named M87. The disk is the wide, bright spot in the upper left. By measuring the speed at which the gas is swirling, as well as its mass, scientists figured out its unusual behavior was caused by a black hole. An early *Star Trek* episode written by Dorothy Fontana predated the term black hole, but used a similar concept: a black star to sling the *Enterprise* back in time to the 1960s. *(Space Telescope Science Institute)*

Entitled "Tomorrow Is Yesterday," it was written during the series's first season by the inestimable D. C. Fontana. Actually, D. C. called the black hole in the show a "black star," but who's quibbling? She used it to warp space-time so completely that it flung the *Enterprise* back to the late 1960s where our lovely starship was quickly and ironically identified as an unidentified flying object!

The reason Dorothy didn't call her black star a black hole was because the term black hole hadn't yet been coined! The man who did that—Princeton physicist John Wheeler—hadn't publicly used the words until the fall of 1967, some ten months *after* "Tomorrow Is Yesterday" aired January 26. Obviously D. C. was ahead of her time in more ways than one! Her prescience also makes you wonder if her black star didn't somehow inspire Wheeler's black hole after he watched the episode. Probably not. Lawrence Krauss at Case Western says he knows Wheeler. "John," he says, "is too much of a showman to have lifted the concept."

Anyhow, this black star or hole or whatever you want to call it, was so powerful that it gripped the *Enterprise* and flung it back three hundred years. It was *Star Trek*'s first time-travel story.

Black holes are just the kind of cosmic phenomena that can pull this sort of trickery off because they are the baddest, beastliest things in the universe. They eat everything that gets near them.

Put another way a black hole is like an infinitely heavy lead ball sitting on the fabric of space and time. I have this on the good authority of several scientists that we pestered about this question. I can't guarantee that I understand all of this as well as they do, but let me attempt an explanation.

Imagine the universe has been rolled out like a great trampoline, with everything that is in the universe represented as objects that are lying on it—stars, planets, comets, entire galaxies. If an object is heavy (and therefore exhibits a strong gravitational field), then, as Einstein pointed out and others have proven, it tends to bend the space and time around it, the same way a heavy object bends the trampoline that is supporting it. It creates a kind of well. The heavier the object, the deeper the well. If a lighter object comes near a heavier object, it will want to roll in its direction, unless it is moving so fast that it simply blows right by it, escaping its "gravitational" field.

If a heavier object rolls by a light object, the lighter object will fall into its orbit. Heavy objects, all objects in fact, bend space, time, light—the whole electromagnetic grid of the universe—to some degree or another. The more something weighs, the more it bends the currents of light and electromagnetism, and the more it slows time down.

Black holes are odd because they are not really that large, but they are incomprehensibly dense. So dense, so gravitationally overbearing, in fact, that anything that comes within their grip is a goner. Nothing gets away; not stars, not planets, not light, not time itself. They are like great cosmic drains sucking down whatever parsecs of the galaxy happen to be in their vicinity.

In their past lives most black holes once made their livings as huge stars, but eventually even the brightest, biggest stars burn out and collapse upon themselves (just like in Hollywood). Not all stars are big enough to become black holes. Our sun, for example, can't attain black hole-osity. In about five billion years it is more likely to become a mere white dwarf, a burned out star that is a few thousand miles across and extremely heavy and dense. Others become neutron stars, defunct metal cores no more than twenty miles or so wide, but millions of times heavier and more dense than dwarves.

The really colossal stars? Those are the ones that become black holes.

Some black holes may also form from huge clouds of interstellar gas, the kind, you know, that the *Enterprise* always seemed to enter immediately before some monstrous disaster struck. These can be millions or even billions of times more massive than our sun. When these collapse, they become the cosmic equivalent of Arnold Schwarzenegger, though they aren't paid as much. A few have even been "spotted." You can't actually *see* a black hole because no light escapes from it, but astronomers have seen evidence of them. In 1994, for example, the Hubble Space Telescope found a gargantuan black hole at the center of the M87 galaxy (dead boring name for a galaxy). It's estimated mass: two to three *billion* Suns; yet it is so seriously condensed that it is no larger than our solar system.

It is even likely that a massive black hole lies at the center of our own Milky Way (mass about 2.6 million Suns' worth!). Luckily Earth is several gazillion miles away, so we won't be getting swallowed up anytime soon.

So what happens if you get near a black hole and what does any of this have to do with space travel?

Well, according to Stephen Hawking, the tiny, yet irresistibly heavy object at the center of a black hole creates a gravitational funnel that turns it into a cosmic vortex. Hawking should know. He's the world's leading black hole-ologist. In the 1960s he developed a theory, now generally accepted, that a "singularity" must have occurred at the birth of the universe. A singularity is a black hole taken to its logical extension—a "place," a point, where *everything* in the universe (not just a piddling few billion suns) has been condensed into an incalculably small, dense point, infinitely smaller than the period at the end of this sentence.

Every black hole is surrounded by what Hawking calls an "event horizon." This perimeter, he says, is the lip of the vortex, the ultimate point of no return. Once anything—you, for example—hits this "event horizon," there is only one direction in which you are headed, and the results will not be pretty. After you are over the edge, tidal gravitational forces will shred you faster than you can say, "Beam me up, Scotty." In fact, the forces at work beyond this horizon are so powerful that the laws of the universe as we know them are entirely repealed. Sort of the way laws break down when you get beyond the perimeter of certain bar doors in Los Angeles or New York. In fact, black holes in many ways are not part of the normal fabric of space and time. The forces at work in their vicinity are so different that they are really pinched off from reality, as we know it—which is what makes them so damned interesting.

Why am I telling you all of this? Partly because it makes me feel smart (aren't you impressed?), but also because . . . well, just trust me, you have to know it. You'll see. Because black holes, even though they are no longer science fiction, *can* render the improbabilities of science fiction possible.

For example?

For example, time travel.

Read on and prepare to become unstuck in time.

I wouldn't take a bet against the existence of time machines. My opponent might have seen the future and know the answer.

—Stephen Hawking

Time is nature's way of keeping everything from happening all at once.

—John Wheeler

Time present and time past are both perhaps present in time future. And time future contained in time past.

—T. S. Eliot

Listen; there's a hell of a good universe next door: let's go.

—e. e. cummings

4

RIDING THE ARROW OF TIME

I've traveled back and forth through time so much my head hurts. First there was that episode, "Tomorrow Is Yesterday," when the *Enterprise* was slung back to the twentieth century by a black star. And then the time we returned (again) to the 1960s and crossed paths with Gary Seven, a mysterious human sent to save Earth by the inhabitants of another world ("Assignment: Earth"). What a mess *that* was. And of course there is the famous and immensely popular "City on the Edge of Forever," where Spock, McCoy, and I pass through a time portal called the Guardian of Forever. I fall in love with the beautiful Edith Keeler (Joan Collins) and face the distressing fact that I'll risk the Earth's future if I attempt to save her life. Fortunately for the *Enterprise* crew, unfortunately for Edith, I do the right thing, she bites the dust, and the future is again set right. And don't forget *Star Trek IV: The Voyage Home*, one of my favorite *Star Trek* movies (even if Leonard Nimoy *did* direct it). It took us back to San Francisco into the 1980s.

There has been a surfeit of other chronologically challenged sci-fi adventures conceived over the years, of course. Time travel

Let's say we really could jump through a time portal like Kirk, Spock, and McCoy did in the classic *Star Trek* episode "The City on the Edge of Forever." Can we play with history? Kill our own grandfather and wipe ourselves out of existence? Some scientific evidence indicates that if we could travel back in time, anything we did would serve to create the present we live in now. Or it may create other events in a parallel universe.

in science fiction is time-honored, and everyone from H. G. Wells (*The Time Machine*) to Michael J. Fox (*Back to the Future*) to Arnold Schwarzenegger (*The Terminator*) has made hay with it.

But the truth is for years I harbored a bias against time-travel stories. I felt they were a cop-out; a convenient way to throw *Star Trek*'s characters into impossible situations and then pluck us, artificially, back out. I suppose that the whole idea of traveling through time and wreaking havoc with the future by changing past events seemed so impossible to me that somehow time-travel stories made me feel that we weren't quite playing by the rules.

However, since working on this book, I've changed my view. I mean I've always known that time traveling is damned tricky business, but now, after speaking with Kip Thorne, the world's leading time-travel expert, I understand it is even trickier than I thought in the first place.

You're wondering, "What is he talking about? Has he been through the transporter too many times? Is he suffering from warp lag? Hello! Bill! Time travel's not just tricky, *it's impossible!*"

Shush!! I would have thought so too, but it turns out that after decades—centuries, really—of we humans conceiving all sorts of wild confabulations about fictional journeys through time, that a handful of scientists may have actually discovered some ways to pull it off!

Time Tunnel Vision

Kip Thorne is the man we sought out on this question because he's the man who made time travel respectable in the scientific world. Before Thorne and two of his doctoral students, Michael Morris and Ulvi Yurtsever, published a paper in *Physical Review Letters* about slipping the bonds of the present to visit the past, time travel, at least in serious scientific circles, was considered the exclusive province of science fiction fanatics, writers, and the mentally deranged (there's a difference?). Time-travel theories made for interesting thought experiments and heavy conversation, but other than that the scientific community dismissed it all as mumbo jumbo.

Thorne, however, was able to tackle it without destroying his career because he is not your garden-variety theoretical physicist. He's one of the world's most respected theoretical physicists, and a leading expert on Einstein's general theory of relativity. He hobknobs with Stephen Hawking; his teacher was John Wheeler (the man who coined the term black hole); and he is the Feynman Professor of Physics at the California Institute of Technology, a West Coast school of impeccable reputation.* So what's a top-flight theoretical physicist doing mucking around in subjects as verboten as time travel? Well, blame astronomer and author Carl Sagan. In addition to the other scientific luminaries he associated with, Kip was also a good pal of Sagan's. It was Sagan who got the whole time-travel ball rolling back in 1985 when he ran into a unique cosmic travel problem that he felt Thorne could help him solve.

Sagan had already written several award-winning science books, but the project he was working on when he contacted Kip was a science fiction novel, his first. It was entitled *Contact*, the story of a smart, gritty radio astronomer named Ellie Arroway who has the immense good fortune to tune into the communications of an advanced, alien civilization light-years away from Earth—rather a pivotal event in human history. After a lot of hard work, she discovers that the extraterrestrials have not simply made contact, but beamed the instructions for building an extremely advanced machine designed to carry a small group of intrepid explorers across the galaxy to their planet.† Basically they were saying, "If you build it, you can come."

Given the premise of the novel, Sagan needed an extremely fast, yet believable way to get around the galaxy, just as *Star Trek* did. Except in this case, an entire starship was not required, just a nifty invention that moved the story's heroine to the Vega star system twenty-six light-years away where the extraterrestrials

*Feynman, a former Cal Tech professor, is considered one of the truly great theoretical physicists (and raconteurs) of the twentieth century. He won the Nobel prize in 1965 for his work in light, radio, electricity, and magnetism. He was a man who worked on everything from the Manhattan Project to the investigation of the *Challenger* shuttle disaster.

†In the novel the trip is made by several astronauts. In the movie only Ellie makes the journey.

lived. This meant, as with all interstellar journeys, that the trip was doomed to be a very, very long one, even if Sagan could get his heroine traveling at near-light speed. And if he *did* get her traveling that fast, he'd still have to deal with all of those energy and time dilation issues again.

"That was my problem," Sagan said in an interview with the BBC before his unfortunate death in 1996. "[I needed] to get her to a great distance away from the Earth in the Milky Way galaxy to meet the extraterrestrials, and then come back, and do all that within the lifetime of the people she [had] left behind."

Originally, Sagan intended to somehow use a black hole to pitch his heroine through the galaxy, but something about that idea didn't feel right. And so, just as Gene Roddenberry often did, he consulted an expert, and that expert was Kip.

Kip listened to Carl and quickly set him straight on the black hole idea. He told him that anyone who planned to travel the galaxy using a black hole was in for a rude awakening: he would be thoroughly whipped and blended the moment he entered. That, he told Sagan, might bring an abrupt and unsatisfactory end to the novel. But Kip didn't simply shoot the black hole approach down. Being the good and knowledgeable pal that he was, he offered a solution. He suggested to Sagan that he explore using not *black* holes, but *worm*holes, and to help, he sent Sagan about fifty lines of mathematical equations explaining how.

Commented Sagan, dryly, "[It was] a level of detail . . . that I had not anticipated."

Wormholes, in some ways, are extremely similar to black holes, except with one very important difference—they have exits. Put another way, at least in a way that I can understand it, they are perforations in the fabric of the universe that tunnel between different sectors of space and time. Theoretically you could enter one end here, and come out the other end in an entirely different part of the universe, just like that!

The idea of wormholes wasn't all new. It had been around since the 1950s when the ubiquitous John Wheeler first discussed—and named—them. (Wheeler obviously has a knack for coming up with memorable names for odd cosmic phenomena.) Despite Wheeler's interest in wormholes, they weren't exactly a hot topic even in a realm as exotic as theoretical physics. In fact they were so little known they hadn't even come up much in science fiction literature. Really they were nothing more than pure conjecture, with one important feature, Kip told me. They were conjecture that had been predicted by Einstein's equations. And that gave them instant credibility.

Traveling Through the Big Cosmic Apple

I now understood that wormholes *could* exist. But *do* they exist, and if they do, how, I wondered, do you use them to jump around something as big as a galaxy? Here's how it was explained. Imagine that the universe is curved like the surface of a gargantuan apple. And further imagine that you're a worm that lives on this apple. Because you're small in comparison to the rest of the apple, its surface looks flat. (Just as Earth's surface

If we *could* create wormholes, ships like this may someday travel through them. And if Kip Thorne's recipe holds, they may even travel through time. *(NASA Glenn Research Center)*

looks flat to all of us.) But let's say that you're a very clever worm (naturally) and eventually you realize that you could make your way to the other side of the apple if you just keep crawling ahead in a straight line. That proves the apple is curved, right? Otherwise how could you ever come all the way around to the other side?

Okay, I get that. And now that I have I quickly see—being a genius among worms—that you can also get to the other side of the apple simply by tunneling directly through it (something worms happen to be very good at doing with apples). By going through this wormhole I have just cut the distance I have to travel to a fraction of what it would be if I crawled across the surface.

So when it comes to cosmic wormholes, we are like the worm, and the galaxy is like the apple—curved. We can't see the curvature because we're so small in comparison to the rest of the cosmos, but experiments have proven when light travels across a large enough segment of the universe, it arcs. Scientists believe that it would only bend like this if the space and time through which it is moving were curved as well. Astounding as this may seem to a pea-brain like me, this means that the fabric of the universe may be spherical. However, it may also be much messier than that. It may fold in, around, and upon itself in all sorts of ways we do not yet understand. But let's be gluttons for punishment and try to understand it anyhow.

Imagine space and time as a very long, flat sheet of paper. Now mark two points on the paper, about a mile apart, and lay it flat on an equally long table. Now you could travel between those two points simply by dragging the pencil from one dot to the other. But there is another, much faster way to travel between the two points. Fold the paper so that one point lies directly underneath the other one, *and then drive the point of your pencil from the dot on top through to the one on the bottom.* Just like that you've made the journey and connected the dots!

In this universe, it turns out that the shortest distance between two points isn't a straight line, as Euclid said, it's a wormhole! And that is precisely what Kip Thorne suggested that Carl Sagan use in his novel to move Ellie Arroway across twenty-six light-years of space. It was fast, credible, and different.

Sagan, of course, loved Kip's solution, and he used it. Ellie made the journey and the book went on to become a bestseller. The movie didn't do too badly either.

What does this have to do with time travel? Well, hang on, because the story doesn't end here. Now that Thorne and his students had started exploring faster-than-light-travel, the three of them just couldn't seem to leave wormholes alone. And that's when something amazing occurred to them: They hadn't simply conceived of a cosmic rapid transit system, they had discovered a time machine!

A Recipe That Defies Time

How, you ask, do you get from wormholing to time tunneling? How does a thing designed to shorten long distances into short ones also turn out to be a machine that whisks you through time? No big leap if you're one of the world's leading theoretical physicists, maybe, but one giant leap for the rest of us mortals. To try to work out a way to make all of this more understandable, Chip and I worked with Kip to cook up (literally) a way to make his wormhole/time machine idea more palatable for those of us who don't have Ph.D.'s in physics. Here's the recipe we came up with. (Look out, Martha Stewart.)

TIME-TRAVEL SOUFFLÉ

This is a timeless little dish that will take you back to the good old days! (Secret tip: It's a little simpler to prepare if you are the member of an arbitrarily advanced extraterrestrial civilization.)

INGREDIENTS:

1 very large universe—curved

1 wormhole with 2 fresh, artificially grown wormhole mouths, linked

1 intrepid chrononaut

Unlimited amounts of "exotic matter"

1 near-light speed engine

SERVINGS:

Unlimited (once you've made the first batch)

PREPARATION TIME:

Depends on how arbitrarily advanced you are. Probably longer than a Sunday afternoon.

DIRECTIONS:

Place the two wormhole mouths near one another in space; make sure that the end of each one is connected to the other. Carefully stir large amounts of exotic matter into the end of one wormhole (we'll call it wormhole doorway #1) until it has reached the center of the tunnel between each mouth. This matter ensures that the tunnel will stay open and remain connected to the other wormhole mouth (doorway #2).

Next rev up your handy hyperfast ship, hook it to wormhole doorway #1, and drag it at near-light speed as far as you can in three and a half days. Then turn around and carefully return the hole to its original location a total of seven days later. Stir in more exotic matter if needed.

Because time slows down when you approach the speed of light, the first wormhole doorway now exists several days in the future as far as the one that never moved (#2) is concerned. In other words the mouth of each wormhole is now in a different time, one in the present and one in the future.

Next place intrepid chrononaut in wormhole doorway #1 and send her on to wormhole doorway #2. When she exits, she will have traveled backward in time!

Voilà! You've just prepared your first time-travel machine!

Serve over baked chicken with a creamy wine sauce. Goes well with a crisp, dry Pinot Grigio.

Now how easy is that? Piece of cake, right? Not exactly. (Apparently no lunch is free, even in the weird world of theoretical physics.) Did you notice the secret ingredient? Something Thorne calls exotic matter or negative energy? (Remember we also came across this with Millis and Alcubierre's starships.) It's this exotic matter that could make the wormhole—in Kip Thorne's language—traversable. In other words no exotic, negative matter, no time travel. No warp drive. What kind of universe is that?

Naturally at this point, I am, once again, confused. To me exotic matter is something sold by a street peddler in Hong Kong. And negative energy? Well, I've come across plenty of it on sets where really bad movies are being shot, but when it comes to understanding what these things could possibly have to do with time travel, I'm in the corner of the class wearing a dunce cap.

Luckily, Kip is a patient, long-suffering man and he explained everything. Einstein's equations not only predicted wormholes, he told me, they also predicted that they would be *very* unstable. They would come and go, open and close, snap into existence and right back out in the blink of an eye. If you did manage to dive into one, the very act of diving, said Einstein's mathematics, would slam the door shut behind you, pinch the hole off from the rest of the universe and, well . . . let's just say your flight will have been canceled.

Since energy is what makes the wormholes so fidgety in the first place, the negative energy could directly counteract it; calm everything down and keep the tunnel stabilized, like cosmic Valium. Apparently, making time tunnels that work requires lots of this stuff.

So how do we get our hands on it? For an arbitrarily advanced (read god-like) civilization, it's not a problem. Just pick a passel up at the local Wal-Mart (right next to paints and wallpa-

EXOTIC MATTER OF FACT

Is negative energy or exotic matter or whatever you want to call it even possible? Yes. Back in the 1940s a Dutch physicist by the name of Hendrick Casimir had an interesting idea. He knew that quantum physics had already postulated that energy can appear out of nowhere, as long as it quickly disappeared again. (Don't you love quantum physics? It's so *goofy!)* All of space, it seems, even vacuums, are ceaselessly boiling with particles and antiparticles, including photons, that pop into existence one instant and then quickly annihilate one another out of existence the next.

If this is the case, thought Casimir, then you should be able to place two electrically charged metal plates close together. I mean really close; so close that you begin forcing certain photons whose wavelengths are too big to fit between the plates into the space surrounding them. As fewer and fewer photons are left between the plates, more and more run around the outside of the plates and start to push from the outside in and that causes them to begin to collapse on one another. Or put another way, the lack of energy between the two plates would begin to draw them together. This is a little like the quantum version of a vacuum seal. (Casimir couldn't prove his theory in 1948, but since then others have.)

What is the force that is pulling the two plates together? If you guessed negative energy, go directly to the head of the class. This means that Kip Thorne's negative energy, a.k.a. exotic matter, *does* exist and could be used to thread wormholes which would in turn make his time machine possible, theoretically. The bad news is it can be manufactured in only very small amounts. At least right now.

And you thought antimatter was hard to come by.

per). If you're just a poor mortal slob like we *Homo sapiens*, it's a little tougher. But not impossible. Negative energy actually does exist.

Does this mean that all we have to do is learn how to manufacture lots of negative energy? No. The cosmic jury is still out on this question, says Kip. "It's true that negative energy can exist, but the laws of physics might also prevent us from collecting enough negative energy into the narrow throat of a wormhole to keep it open. In fact recent research seems to indicate exactly that which would mean that nothing will ever be able to travel through a wormhole. I'm not sure if that is true or not. I have my doubts. Either way we have to probe the laws of physics a lot more before we know the answer for certain.

"But let's say you *can* use negative energy to hold the throat open, it's still not a sure thing that the wormhole can absolutely be used as a time machine. My own most recent research indicates that if you try to activate a time machine, any kind of time machine, it'll self-destruct.* Frankly both Stephen [Hawking] and I are pretty pessimistic about the possibility of time machines, though I still think that wormholes exist."

The Grandfather Paradox

Personally I hope time machines *are* feasible. The possibilities are just too intriguing. So just for the sake of argument let's suppose that we have oodles of exotic matter and that the matter can be used to stabilize a wormhole and that we can create a time machine using our handy recipe. One end of the wormhole sits comfortably in the wall of your living room. The other exits in the next room one year before you were born. You're all ready

---

*For more details on why Kip has come to this conclusion, see the last chapter of his book *Black Holes and Time Warps: Einstein's Outrageous Legacy*.

to roll, except now you're faced with a new challenge—time paradoxes.

I will explain to you what has been explained to me. (Then maybe you can explain it back and I'll actually understand it.)

Let's say you want to see what your parents were like before they brought you into the world. Or you want to buy Intel or IBM at $1.50 a share. Or you want to understand why everyone wore all of those horrible clothes in the 1970s. Whatever. You step into the wormhole like a modern Alice and zap! out you emerge into the past. As you are wandering around, you bump into your father, literally, and knock him in front of a bus. Wham! No more dad. Now what? You just killed your father! If that's the case how can you be standing there? No father, no marriage, no sex, no you.

This is what's known as the "grandfather paradox," and it is a big problem. It's called the "grandfather paradox" because of a question that Hugo Gernsback, the legendary science fiction editor in the 1920s and 1930s of *Amazing Stories* and *Science Wonder Stories*, once put to his readers in the December 1929 issue of *Science Wonder Stories*. In an open letter he basically asked them if a traveler from the future could possibly interact with people and events in the past. He put it this way . . .

> Suppose I can travel back into time, let me say 200 years. And I visit the homestead of my great-grandfather, and am able to take part in the life of his time. I am thus enabled to shoot him while he is still a young man and as yet unmarried. From this it will be noted that I have prevented my own birth, because the line of propagation would have ceased right there. Consequently, it would seem that the idea of traveling into a past where the time traveler can freely participate . . . becomes an absurdity.

This turns out to be a serious scientific issue, and it is only one of several weird "what ifs" that can entangle time travelers.

What if, for example, you live in the future, and send a time machine back to a time earlier in your life? Then suppose that earlier in your life you travel to the future so you can send the machine back to yourself. So who invented the machine? Your future self or your past self?

Or what if you travel into the future and see yourself tortured and killed in some horrible way, like by having to talk to an insurance salesman for a whole evening? To avoid such a horrible fate, you return to the past and peacefully commit suicide. But how could you die now, having seen yourself die in the future? (And more important, would anyone collect on the insurance?)

Anyhow, you get the idea—moving back and forth through time would seem to make the very act itself impossible. The messiest example of this in all of science fiction was illustrated in a terrifically weird and creative story written by Robert Heinlein more than fifty years ago. Entitled "All You Zombies," it's the story a single time traveler named Jane who manages to pull off the nifty trick of becoming her own mother, father, son, and daughter. Talk about a paradox!

The story goes like this. A baby girl is mysteriously dropped off at an orphanage in Cleveland, Ohio in 1945. She grows up, alone and sad because she has no idea who her parents are or why she was abandoned. If only she knew! When she's eighteen she meets a drifter and finds him strangely attractive. They fall in love, and life is looking good. Then all hell breaks loose. First she finds she's pregnant with the drifter's child. Then the mysterious vagrant disappears. Next, during a complicated delivery, doctors discover she has both sets of sexual organs and to save her and the baby, they transform her from a woman into a man. And then, as if all of this isn't enough, someone, for no apparent reason, kidnaps her baby from the delivery room.

Broken and confused, "Jane" now becomes a bum. For years

he wanders the country until in 1970 he stumbles into a bar and tells his pathetic story to an old bartender. The bartender offers a way for Jane to avenge the good-for-nothing cad who got her pregnant and left her abandoned. However, first he has to agree to join the "Time Travel Corps!" Our hero/heroine agrees and the next thing he knows he and the bartender have entered a time-travel machine and returned to 1963 where the bartender drops him off. Guess who the drifter runs into? An eighteen-year-old orphan he finds strangely attractive and with whom he fathers a child.

While that's happening the bartender jumps into the time machine, jets nine months into the future and steals Jane's new-born baby, and then heads back nineteen years (minus three months) to 1945 where he deposits the infant at an orphanage in Cleveland. *Then* he goes back to the 1960s, picks up the drifter—more confused now than ever—and leaves him in 1985 to join the Time Travel Corps. Eventually the drifter recovers from all of this insanity and becomes a respected member of the Corps. However, his strangest mission still awaits him. He must disguise himself as a bartender and arrange to meet a strange and dejected drifter in 1970.

So who is who here? Who are Jane's mother and father? Who is Jane's lover? Who is Jane's baby? Who is the bartender? Who, for gods sakes, is Jane?! The answer is, they *all* are, except they all came out of nowhere. The whole strange story and the "family" that evolves out of it is one long and twisted loop that's entertaining as hell, but totally senseless.

It's these kinds of illogical paradoxes that have made physicists insist that there is simply no way that time travel is possible. It violates the basic scientific principles of cause and effect. If you could actually perform this sort of mischief, then nothing in the past or future or present would ever make sense because it would all be subject to random revision. Someone goes back in time and prevents John Kennedy from being murdered in 1963 (a favorite story idea of Gene Roddenberry's). Hitler is assassinated. Einstein loses his notes before publishing his *Theory of Special Relativity*. Winston Churchill dies in the Boer War. Gene Roddenberry is prevented from selling a certain series to NBC. Obviously it can't be possible to play with past and future events as if they were blocks in a child's playpen, or life would make even less sense than it already does. We'd all be popping in and out of existence like bad special effects.

Time paradoxes really bother Stephen Hawking. He keeps wondering why, if time travel is possible, tourists haven't overrun us from the future. There are different theories. Some say that there's a lot of time to travel through. Eternity covers a lot of ground. Maybe chrononauts just haven't got around to visiting us yet. Another theory is that maybe time travelers *are* here, but we can't see them because they would rather not be seen. (After all they are arbitrarily advanced, right?) Or perhaps we see them as ghosts or unidentified flying objects. But then says Hawking, "I think that if people from the future were going to show themselves they would do so in a more obvious way. What would be the point of revealing themselves only to cranks and weirdos who wouldn't be believed?" (Maybe because they wouldn't be believed?)

Senselessness

Does all of this chronological angst mean that time travel is impossible? Are studio lot special effects the only way we're ever going to accomplish it? Well, maybe not, or so I'm told. For one thing, the more scientists dig into the nature of the universe, using Einstein's theories and field equations as picks and shovels, the more they realize that even things that look apparently true often don't make much sense.

TIME MACHINES MADE EASY (DON'T TRY THIS AT HOME)

If we can use rock solid mathematics to eliminate all of the reasons why we can't travel back in time, what's to stop us from taking a trip? How about monumental engineering problems. It's one thing to say it's *not impossible* to tunnel through time, another to actually build the tunnel.

Kip Thorne himself, however, has addressed some of the problems. Thorne's exotic matter, the negative energy that we need to keep both throats of a wormhole open, seems to be the magic ingredient. But, again, just because you fervently hope for negative energy to exist doesn't mean it does. But here's a pleasant surprise. It's for real. Remember the Casimir effect? Back in 1948 when Hendrick Casimir theorized you could create negative energy, he couldn't prove his theory was correct because there was no way at that time to get his two metal plates close enough together (billionths of a meter). Now there is, and in more than one laboratory, Casimir's equations have proven to be right on the money. Negative energy not only exists, but can be produced . . . but only in vanishingly small amounts.

But if it can be produced in small amounts, maybe someday it can be produced in large ones. All we have to do is wait until we become arbitrarily advanced. (Based on, say, the popularity of Jerry Springer, I don't think we're there yet.)

But for fun let's assume we are, and we can not only make exotic matter, we can make great heaps of it and shape it into a cylinder big enough to hold you. Go ahead, stand inside the cylinder. Outside the cylinder, negative energy shreds space and time and wraps it around itself linking it to another section of the fabric. Imagine it as the time version of a phone booth, and you're making a collect call to the past. When the call is connected, you simply walk through the cylinder and come out the other end elsewhere in a different time.

Of course to make these time-booths useful, scientists imagine that we'd have to string them up and down the space-time continuum so that they're easily accessible, and since building one would require the use of one in the first place, we might have ourselves a cosmic catch-22. But one thing at a time. For now, this is as close as we're going to get to the Guardian of Forever.

What?

Hang in here with me, and let me see if I can explain to you what has been explained to me. Newton's equations described a universe that was consistent with our "senses." It was literally a "sensible" universe. Einstein's don't. Does it make sense that as you approach the speed of light time slows down, and you grow smaller and your mass increases to the point of being infinitely heavy? No. Yet lots of experiments and observations and mathematical equations have proven those very statements to be absolutely true. They are senseless because they fall outside of the ability of our senses to experience them. But just because we can't see and hear and feel and smell them doesn't mean they aren't true.

Or think of it this way. Let's say you're a tribesman born in the rain forests of the Amazon jungle. One day I walk into your village and tell you that the box I am aiming at you is capturing your image, sending it up into the sky where it is being bounced back to millions of other people around the world (people that you are not even aware exist). Furthermore, because I am "capturing" you, they can see and hear what you are doing right now, right here, even though they are somewhere else far away.

Now since I can't prove any of what I just told you, my guess is your reaction will be to laugh yourself silly. You would call together all of the other villagers and point at me, shaking your head: "Poor, crazy son of a Canadian. The heat and humidity are too much for him. He's lost his mind." And from your point of view, that's a sensible conclusion because what I have described to you is literally beyond your experience. But that doesn't mean I am wrong. We beam television signals all around the planet to and from satellites to billions of people twenty-four hours a day, seven days a week. And we've been doing it for quite some time; since before *Star Trek* went on the air, in fact.

So as far as time travel is concerned we are living deep in the equivalent of the rain forest. Much of what is going on in the universe is obviously well beyond the puny bandwidth that our minds can muster; probably with good reason. We didn't need to know about the inner workings of the universe to survive on the savanna, we just needed to know enough to get a meal, a good night's sleep, and a healthy mate with whom we could copulate. Evolution gave us the tools necessary to continue the species, it didn't give us the tools (at least not at birth) to figure out the elusive mysteries of the cosmos.

The point is that even the seemingly senseless may in fact be true, including something as daffy as time travel. But what, I ask, haplessly and utterly confused, about all of those time-travel paradoxes? We *don't* live in a totally chaotic universe where events change at random as though someone were sitting on a cosmic couch blithely channel surfing the time and space around us.

So which is it? Can we travel back in time or not?

There may be an explanation that both eliminates paradoxes *and* allows for time travel. That's the good news, I think. The bad news is that to really understand how this can be, we have to start wrestling with . . . quantum physics. This is the simplest way I know how to put the premise, based on what Kip and others have told me. It may be possible to travel back in time *and* interact with others *and* even effect the shape of the future. How? Because whatever you do on your travels, you cannot derail the course of the future from which you came. You are too busy *fulfilling* it.

Remember in "City on the Edge of Forever," I wanted to save Edith Keeler. If I had succeeded, Germany would have won World War II and all of world history would have changed. But, McCoy (who was the whole reason we had gone back in time in the first place) tries to save her and I stop him. That very act sets the course of events so that they end up shaping the future that we are all familiar with (except that by the time we're familiar with it, of course, it's the past).

What this means is that even if your intention is to change

the future, you will either be prevented from changing it, or in the very act of changing it, you will set into motion exactly the events you wanted to change in the first place!

Understand? If you do, you've been spending entirely too much time alone reading science fiction.

This turns our normal perception of time on its head because generally we assume that what we do in the present sets up the events of the future. We see cause happening in the past and effect happening in the future. But, clearly, what I have just described indicates that the future, on some level, actually shapes the past! If an event that takes place in the future insures that all acts in the past (whatever they are) must lead to the fulfillment of that event, then the future has guided past events, right?

This is bull, you say. The future doesn't control me. I control the future, at least my own future. I don't believe in predestination. The decisions I make now shape what lies ahead.

Well again, I am told (and therefore I pass along to you), don't assume that what is true is only what you see because you may not be seeing the whole picture. Maybe, if time and space, are both cut from the same cosmic cloth, time has a geography just as space does. Maybe future events are already laid out as surely as the Rocky Mountains were laid out before pioneers headed up the Oregon Trail. We may not have known what lay ahead, but it wasn't because it wasn't already there. If they had a helicopter and could have gotten off the ground, gained a new perspective, then they would have seen the journey that lay before them in an entirely new light.

Can this really be true?

Paradox in the Corner Pocket

Kip Thorne explains it this way. We're playing billiards. Just a regular game that doesn't involve any time machines. In this kind of game the interactions of the billiard balls on the table are all pretty simple. They whack into one another and carom in various directions each according to Newton's laws of physics. This kind of game represents our perception of the universe before Einstein came along.

Now let's assume we *do* have a time machine, and only one billiard ball sits on the table. And let's say I send that ball into the mouth of a wormhole—corner pocket. After a second, the ball shoots out of the exit end of the wormhole right back onto the table just before the original ball enters the front end (remember it has used the wormhole to travel backward in time). It then slams its earlier self sideways so that it can't go into the corner pocket we shot it into in the first place. Impossible, right? A ball can't knock itself out of its own way. This is the billiard ball version of going back and killing your grandfather—a paradox.

At least Kip, and others, thought so. So next they set out to discover a scientific resolution to this problem. One of the others was a Russian scientist and friend of Thorne's, Igor Novikov, who started playing around with Einstein's equations to see if they applied to actions like changing the future by having a billiard ball knock itself out of the way in the past. They didn't. "Einstein's equations were not designed to handle time-travel paradoxes," says Kip. "But Igor realized that with a little work they *could* be made to deal with paradoxes if you demanded that the solutions to the equations be self-consistent. In effect, he speculated that self-consistency was a new principle in physical law that made it possible for a universe to exist where time travel was possible. This doesn't mean that it is, but a lot of theoretical work in recent years seems to indicate that he's right." If you're having a hard time grasping this, I'm right there with you. But after many clarifications the long and short of it seems to be this: *the rules and regulations of the universe will not allow you to rearrange the future by dropping in on the past.*

SPOOKY ACTIONS

Up for a little more quantum thinking? Too bad if you're not. If I have to do it so do you. Here's another experiment that explores the weirdness of time travel at the atomic level. Hold onto your thinking cap.

Based on their studies of James Maxwell's equations (the ones that unified the worlds of electricity and magnetism), two scientists—an American by the name of John Cramer, and an Australian named Huw Price—have separately developed similar quantum theories that deal with the ass-backward concept of the future shaping the past.

Let's say a photon (an electron of light) is headed across your desk toward a piece of dense black paper. Let's also say the paper has two tiny holes in it that are very close together. Which hole will the photon pass through? Price and Cramer say that the photon (again based on the proven equations of James Maxwell) sends out an "offer wave" ahead of itself, a sort of feeler. The offer wave interacts with a particle ahead of itself (essentially a particle that exists in the future). The offer wave is out there saying, "Yo! Got a photon coming! Heads up!" And this particle sends back an "echo" from the offer wave that travels through only one hole in the sheet of paper. And guess what, the hole through which the echo returned, *that* is the hole that the photon of light will pass through, every time. In other words, an echo from the future has indicated which hole the light will pass through in the present. Or, put another way, the future is affecting the past, relatively speaking.

Einstein, in a monumentally nonmathematical statement, called this "spooky action at a distance." Obviously he had a way with words as well as with numbers.

Why am I putting you through all of this? Because it mathematically illustrates that the arrow of time may not fly in only one direction. The future and the past might be communicating with one another in *both* directions on a very fundamental level. And if the past and future somehow communicate on this very basic level, then maybe the same "spooky action" (see above) is happening on our level as well, except we don't see it that way because our senses don't perceive it. It literally "makes no sense!"

But lest we drown in quantum minutiae, remember the broader issue we are trying to explore here: Time travel, and the possibility thereof. These "spooky actions" and "caromed billiard balls" seem to mean that time travel could well be possible for the reason that no matter what you do when you go backward or forward, you can't change the course of the future.

But if that's the case, does it mean you don't have free will, or that the future has already been decided? Does it mean we're all just a bunch of poor slobs who *think* we're in charge when in fact we're really a part of some elaborate cosmic television episode where the ending has already been decided? I hope not.

But maybe it doesn't have to be one or the other. Maybe it's not a choice between trampling through time and wreaking havoc with the cosmos, or being chained to some immutable series of preordained events. Maybe our time-traveling abilities fall somewhere in between. Perhaps we have free will, but within limits. Novikov puts it this way:

"I can have free will to walk along this wall here without special equipment. It's my free will. But can I do it? No, I can't. Why? Because of the law of gravity. It's forbidden."

I can live with that. Maybe we *can* make decisions on our own, but we can only *act* on those that don't defy the laws of the universe. We can't do *everything* our little hearts desire. We never could. This means that though it might be personally convenient for me to go back in time and arrange to win the lottery, doing so would violate the current by-laws of the universe. So the good news is, I can go back (once I build my time machine). The bad news is I just can't willy-nilly change the course of cosmic events during the journey.

So build your time machine and get hoppin'. Won't Gene Roddenberry be surprised when you drop by at the story meeting for "City on the Edge of Forever?" Just remember, you can't change the plot. But do give my best to Joan Collins. And if you get the chance, you might mention to myself that thirty-five years later, I'm still alive and having fun.

Guy's pulled over for driving under the influence. When he's hauled into night court, the judge slaps him with a stiff fine and sentences him to thirty days in the local hoosegow. Throughout the proceedings the man stands before the judge, weaving noticeably, head hung in shame. Finally the judge looks down upon him from high behind his bench and says, "Now, young man, have you anything to say for yourself?"

The guy looks up, still clearly feeling no pain. "Just one request, your honor."

"Yes?" replies the judge.

The man pulls his wallet out of his back pocket, flips it open, and says, "Beam me up, Scotty."

—*Story making the rounds for years at* Star Trek *conventions*

I like you, but I wouldn't want to see you working with subatomic particles.

—*Bumper sticker*

5

SCRAMBLED ATOMS

Who among us couldn't have used a transporter from time to time? I know there's been an occasion here and there in my life when being "beamed up" would have come in handy. But . . . that's another book.

Star Trek has certainly found the transporter useful over the years. It is unquestionably one of the most creative of the series' inventions, yet its creation was hardly inspired by creativity.

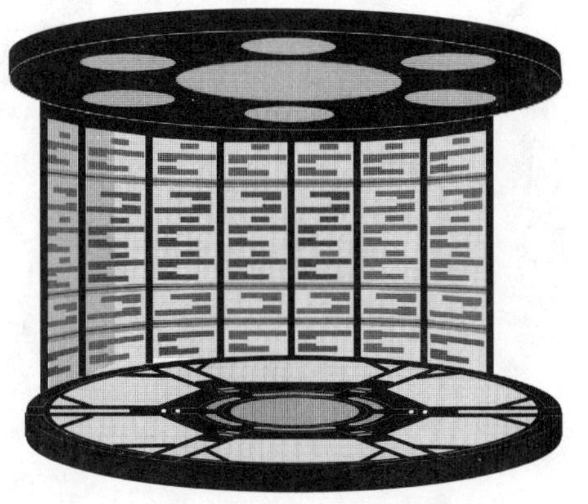

The atom scrambler. Beaming up could be dangerous to your health. Would it be better to beam the information or the atoms? *(Doug Drexler)*

It turns out that Gene Roddenberry conceived the transporter because he abhorred the idea of landing cigar-shaped rockets on alien planets. Prior to *Star Trek* this was the obligatory way that humans arrived on other worlds. There was always a fiery descent, and a lot of false drama as the intrepid explorers emerged from the winged cigar and proceeded to get themselves zapped or manhandled by various aliens. Gene hated clichés and knew that anyone who watched one of those wobbly rockets set down would instantly realize that it was a cheesy special effect.

But if we didn't *land* on the surface of a planet, how were we supposed to get there? Our mission, after all, was to "explore strange new worlds." We couldn't do that by circling an alien world and ogling it through a pair of binoculars. "Hey, Spock, look at this. That looks cool. Doesn't that look cool?"

Thus . . . the transporter, which eliminated *any* need for the goofy retrorocket landings, was born. It also saved a lot of money. Our budget was tight, tight, tight.

Personally, I don't remember anything about the origin of the transporter. I don't recall someone coming up with the term, or a precise moment when the idea hatched. It almost certainly was conceived before the character of James Kirk was. So to try to get the inside skinny, I once again consulted with D. C. Fontana.

"It was just an absolute necessity," Dorothy told me. "We knew we weren't

We're ready to beam. Nothing quite like the transporter had ever been seen on television prior to *Star Trek*. The phrase "Beam me up" will forever be a part of the world's vocabulary.

going to land the ship on a planet. Gene didn't want that. Plus the design of the *Enterprise* didn't make it feasible. She wasn't really landable. So he posited that the *Enterprise* was built in space. She lives in space. She dies in space. Well, given that, the only way you were going to get the crew to the planet, and back, was to beam them there."*

"But where did it come from?" I asked. "Entirely out of Gene's head? Did he suddenly have an insight in the shower? Had some other science fiction writer thought it up first?"

"Well," D. C. told me, "there had been science fiction concepts where characters would step into something like a phone booth. You know, you walk into a contraption in Boston, program your destination and, *zoom*, step out in Tokyo. So the idea of transporting a body in some form somewhere wasn't entirely new. But we tried to give it a new look and a greater air of reality."

"How?"

"Well, once Gene had settled on the transporter concept, the next problem was to work out exactly how it would look and operate. Gene probably discussed all of that with the series's art directors, Pato Guzman and Franz Bachelin and, finally, Matt Jefferies. The single phone booth concept was chucked immediately. Instead we came up with the platform because it was visually more dramatic. You step up to it or you step down off it, like a stage, which drew attention to the whole process.

"Our transporter also allowed you to beam more than one person—both a practical and visual issue. There was an inherent drama in watching people come and go, appear and

---

*I later learned that Matt Jefferies did actually design "landability" into the *Enterprise*, just in case. Retractable feet were conceived and folded up into the saucer-shaped section of the ship, but they were never used.

disappear. They put those translucent, moiré panels behind the platform because they had an interesting, almost moving pattern when you lit them. And then you have the cylinders up at the top and the light from the bottom, which helped the "sparkle" effect you saw as a body appeared or disappeared."

The transporter had other practical advantages too. As Gene liked to say, it made it possible to get right into the story, "by script page two." Think about the way *Star Trek* episodes traditionally opened.

Enterprise orbiting a planet. Voice over. The captain's log entry sets the stage for the story and, *bam!*, there we are appearing on the planet as a bunch of congealing atoms. No wasted time, we were right into the adventure; the plot tease was completed, go to commercial break and opening credits.

That explains why we used the transporter on the show, but can you really create one? I sure hope so. I mean I want a transporter, and I want it *now!* I want them to replace planes, trains, and automobiles. I want a chicken in every pot and a transporter in every garage. Think of the hijackings and automobile accidents we'll eliminate. Think of all of the babies that will be spared wailing as jets descend for a landing. Okay, it will be more difficult to accumulate frequent flier miles, but I'm willing to make sacrifices. But do the laws of physics care what I want? Is there really any chance that science can create a machine that can beam you directly to your seat at the baseball stadium, or your Martian vacation home?

If you did a Vulcan mind-meld with me, I couldn't tell you.

Melting Atoms

Lawrence Krauss's university office at Case Western Reserve University isn't what you'd expect the office of the chairman of the Physics Department to look like; you know: refined, mahogany desk, book-lined walls with busts of Isaac Newton, James Maxwell, and Albert Einstein sitting around. Instead Krauss's office looks like a photon torpedo hit it. Books, magazines, papers, toys scattered everywhere. Near one wall stand two cardboard figures, one of Kirk, the other of Spock, circa 1968, looking blankly at whomever enters the room. That was a little spooky.

The shelves of Krauss's desk are a hodgepodge of objects—pictures of his daughter, physics books, a *Star Trek* Ken and Barbie set. On one wall above his computer is a huge poster of Albert Einstein dressed as a muscular, brightly colored superhero knocking out two bad guys.

You gotta like the man's attitude.

It says in the *Star Trek: The Next Generation Technical Manual*, right on page 103, that the transporter works this way: "The molecular imaging scanners derive a realtime quantum-resolution pattern image of the transport subject while the primary energizing coils convert the subject into a subatomically absorbed matter stream . . . The matter stream is briefly held in the pattern buffer, which allows the system to compensate for the Doppler shift between the ship and the transport destination. The pattern buffer also acts as a safety device in case of system malfunction, permitting transport to be absorbed to another chamber."

Sure, easy for Starfleet to say, but what I want to know is do any of those words make sense or is it all just a bunch of pseudo-scientific mumbo jumbo? Could such a thing actually work? And, if so, will it be soon enough for summer vacation?

Answer: Probably not. In fact, definitely not, decidedly not. Why? Because deconstructing human beings and then beaming them elsewhere is, shall we say, just a *bit* energy intensive.

So what, you say, lots of things are energy intensive. Getting out of bed in the morning. Reading tax forms. Talking to your mother-in-law! They're *all* energy intensive. Yes, but they pale when compared with what it would take to transport even a human pinky to the surface of planet Zork. Because when it comes to beaming "subatomically absorbed matter streams" we are talking about burning energy on a *cosmic* scale.

Why, you ask, don't you just scramble the atoms and shoot them down to the surface of the planet and get on with the story. Well, I don't know why it takes so much energy, but Krauss does and he told me that it has to do with the same reason you can't walk through walls.

It seems that when you bang your hand against a table or whack your head on the car roof as you're getting in or out, it's *not* because you and the offending object are actually solid. In fact there is so much space between the molecules that are you, and those that make up the objects around you, that you should be able to pass right through the thickest concrete.

But it's not simply the density of an object that makes it "solid," it's the electrical fields that hold the nucleus and electrons together; it's the force that keeps the universe from splattering in all directions. It takes *a lot* of energy to break these fields apart. (If it were easy, nothing would be stable.)

Nevertheless, says Krauss, it *might* be possible to beam people through the ether if all that *Star Trek*'s transporter had to do was break molecules down. That can be done chemically, with powerful but manageable amounts of energy that we use in industrial situations every day, like dynamite. But that is not what *Star Trek*'s transporter does. When transporting it breaks you apart not at the sub*molecular* level, but at the sub*atomic* level!

Subatomic particles are 100,000 times smaller than atoms and the forces that keep those particles—the protons and neutrons within an atom's nucleus—from flying apart, are millions of times more powerful than the forces that hold molecules together. (This is why governments spend billions to build particle accelerators that are miles in diameter to smash atoms to pieces, and even then manage to break them apart for nothing more than milliseconds.) These connective forces keep the universe from constantly boiling and shredding itself into cosmic scrap. They put the fabric in space and time, and the continue in continuum.

When you try to break *these* constituents apart, not only does it take prodigious amounts of energy, says Krauss, but it also *releases* prodigious amounts of energy. This is why it's generally not a good idea to be around an atomic bomb when its atom is split to create a chain reaction, or why the sun can still burn your skin even though it's fusing helium out of hydrogen atoms 93 million miles away.

Now, on a good day, when I am all there, I consist of roughly 10^{28} (that's a 1 with twenty-eight zeroes behind it!) atoms. Lawrence Krauss told me so. He also told me what it would take to break me down to my most basic parts (quarks); to literally dematerialize me, pack me in a "matter stream," and transport me from, say, my office to my favorite reading chair at home.

"All we have to do," says Krauss, grinning just a little too broadly, "is heat you up to about 1000 billion degrees (about a million times hotter than the core of the sun). This means generating about the same energy that you get when you detonate one hundred one megaton hydrogen bombs."

Ouch!

Given this fact, it's difficult to disagree with Krauss when he says that transporting people around the galaxy just isn't very "environmentally friendly." It's also, at least based on what we know right now, impossible, because even if you could manage to disintegrate someone, how would you gather up all of the quarks in a nicely organized beam, ship them elsewhere, and then reassemble them in the right order when they got there?

Forget it.

Make Information, Not Matter

So Krauss suggests an alternative. Don't beam the *atoms*, beam the *information*.

"I have surfed the Internet and have seen the light," he says. "I know that I can move information a lot faster than I can move matter."

Aha! Now with that, a light goes on in my head. Even *I*, the Luddite, get the Internet analogy. Over the Internet we don't move objects, we move information, little tiny bits of information. If I wanted to send you a copy of this book, I *could* dematerialize all of its atoms (incinerating myself and most of Los Angeles in the process) and beam it to you. Or I could send the digital representation of it as an attached e-mail document over the Internet. (Considerably safer.) One approach sends atoms, the other sends zeros and ones. Which makes more sense to you?

So Krauss proposes modifying Starfleet's transporter system. (He'll have to take this one up with some desk jockey. This isn't my area.) Forget about beaming bodies, beam the information that makes it possible to reassemble your body at the destination. No more subatomic matter streams, no more hydrogen bombs, no more fat utility bills. Just scan the information that represents you, the way a photo is scanned into your computer (more or less). Represent it as a big bushel of zeros and ones and zap it to your destination where the digital information arranges to snatch the requisite molecules from the air, assembles them in the right order, and abracadabra, *you are there!*

It sounds good, except for this small problem: How do you digitally represent a human being? We aren't talking about representing the *image* of a human, or even a general description of one, we are talking about scanning the location and makeup of all 10,000,000,000,000,000,000,000,000,000 atoms that are you.

"There is an enormous amount of raw information in a human being," says Krauss. "I did the kind of ballpark calculation that physicists like to do and was stunned at what I found. Let's say it takes about a page of information to describe the configuration of each atom in your body. You know, where it's located, what its nearest neighbors are, what its energy levels are—all of that stuff. There's probably more, but let's just say that's about right. How much information is that?"

"I'm guessing . . . a lot."

"Good guess," says Krauss. "To represent that information you could fill up enough ten gigabyte hard drives to reach from here to a third of the way to the center to the Milky Way galaxy. (Chip's note: that's about ten thousand light-years or 59 quadrillion miles, give or take a parsec or two.) But just for kicks let's say we solve the problem of storing that information accurately. How long would it take to send all that information from a ship to the surface of a planet that it is orbiting? Longer than the present age of the universe, at current transmission rates."

This might create some problems. Imagine McCoy and Kirk standing on the surface of Omega IV, beads of sweat on their noble brows, in a serious hurry to beam up. "Two to beam up,

Scotty. Now!" Scotty stands in the transporter room, but all he can do is watch the little hourglass on his Windows-operated monitor screen and say, "Just as soon as I download this information. You've got to give me more time, Cap'n!"

Kirk (impatient): "How much time?"

Scotty: "About eighteen billion years."

"Kind of kills the dramatic tension," says Krauss.

Yeah.

However, there's an important phrase to remember here, "at current transmission rates." Krauss is talking about downloading and processing digital information at the upper limit of what we can do now. You have to figure, especially with how rapidly computer technology is advancing, that the computers of the twenty-third century are going to be *seriously* faster than anything we have now. When you take that into consideration, it's a different matter. "It's the difference between what is impossible," says Krauss, "and improbable."

Let's assume that computing speed and power will continue advancing, conservatively, at one tenth of its present pace. Krauss calculates that *Star Trek*-ian computers will be 1,000,000,000,000,000,000 (a million trillion) times faster in the twenty-third century than they are now. This would go a long way toward eliminating that hourglass on Scotty's screen. In fact, it would be fast enough to calculate the information needed to rebuild a human within a few seconds.

All we have to do now is figure out what we're calculating in the first place!

Slippery Atoms

Now it gets ugly, I mean really confusing, because you have to think about what a human being is, just on a physical level, forget the metaphysics. Never mind questions like, "If you scan the brain are you also scanning every memory, hope, and dream? Or is scanning a brain and body the same as scanning a soul?"

We won't go there. First, there are other more mundane problems to solve. For example, none of us is static. A book is static, a picture is static, even movies are a rapidly changing progression of static images. But a living thing is constantly changing, it's chaotic. It's not just that our hearts are beating and blood and hormones are coursing through our bodies, but cells are dividing, thoughts are being created, insights manufactured, emotions are erupting. Millions of synapses are firing every second inside of our brains, and billions of chemical-electrical reactions have to be taking place for all of this to happen.

Tell me, how, exactly, do you scan *that*. And then how do you represent it digitally? And, once shipped through the cosmos, how do you rebuild it exactly as it was before somewhere else? The answer is that you have to invent a machine that can scan you all the way down to your quarks.

But, Bill, you gasp, not quarks! Is it really necessary to get *that* detailed?

I'm afraid so. Have I ever steered you wrong before? (Don't answer that.) It seems that this is the only way to be certain that we are going to get a really accurate representation of you. (Would you want it any other way?) Of course, if we get down to that level, we have still another problem. When dealing with things that are this small, you can't measure *anything*.

It's a problem.

Even the most seasoned physicist will tell you that this is a world that *no one understands* because it deals with quantum mechanics. I found this tremendously uplifting because even some of the brains we were interviewing for this book couldn't really explain quantum mechanics to me. I even briefly entertained the thought that maybe I wasn't so dumb after all.

The more I asked my naïve questions about how the universe operates at this level, the more puzzled and frustrated these extremely intelligent people would look. Part of their puzzlement was no doubt due to trying to figure out a way to explain such complex concepts to a man who struggled to remember his lines for *3rd Rock from the Sun*, but I don't think it was just that. In fact I know it wasn't because they would often admit at some point that *they* didn't understand it. The best they can do, they would often finally say, is observe it and describe it. But no one really *understands* it.

Murder by Beaming

The culprit that contributes most to all of this confusion is best explained by what is known as the Heisenberg Uncertainty Principle, an insight originally articulated in 1927 by Werner Heisenberg, a Nobel laureate in physics and a man of unquestionable genius. This is something I'd love to explain to you, but . . . I'm too uncertain. Instead I ask Dr. Krauss.

"The Heisenberg Uncertainty Principle," explains Dr. Krauss, "states you can't scan down to the subatomic level because you can never know exactly where every atom in your body is located at any given moment. Or if you know *where* it is, you can't know exactly what it's doing at the same time. In other words, you can't know its velocity *and* its position simultaneously. I can know one or the other, but the combination always has some uncertainty.

"This means that the person I put together after they've been transported would not be the same person I started out with, at least not at a subatomic level because when I scanned them, I couldn't nail down the exact position and velocity of every atom. Would that be important? Who knows? I'm not a biologist. I don't know if you'll have an eye in the back of your head or just a headache. But I know that all of your atoms wouldn't be in exactly the same place. Maybe it wouldn't be a big problem. On the other hand, memories and that sort of thing are probably encoded on at least an atomic level, so if you don't have all of the memory atoms right, what does it do to your recollection of the past?"

I forget.

The point is even if we can figure out a way to beam digital versions of ourselves, *and* even if we can develop the technology to do it fast enough that it doesn't take forever to accomplish, *and* even if we can master a way to pluck atoms out of the air to remanufacture a body right then and there on Iota 47, we still can't be certain that what gets beamed is, in fact, precisely the "us" that we started out with.

To be honest there are certainly memories I would welcome forgetting, but then there are far more I'd rather hold on to. So I'm not sure I like the random effect beaming could have on me, no matter how much I abhor traveling on airplanes.*

Another issue with Krauss's digital transporter is that once you've scanned a human, what do you do with the original? When you copy a piece of paper you have the original and a copy (or a lot of copies). That's fine with paper, but damned messy with human beings. Using Krauss's transporter, the

---

*How does *Star Trek* handle this whole uncertainty issue? We made something up. It's called a Heisenberg compensator. *Star Trek: The Next Generation Writer's Manual* (not for sale) explains it this way: "From the pattern buffer, the molecular stream and the coded instructions pass through a number of subsystems before reaching the emitter. These include the subspace doppler and Heisenberg compensators. Each works to ensure that the matter stream is being transmitted or received in the correct phase, frequency, and so on." Not much of an explanation really, is it? I like Michael Okuda's answer to a *Time* magazine reporter when he was asked how Heisenberg compensators worked. "Very well, thank you," he said.

moment Spock is scanned and beamed, then you have two Spocks, the one on board who was scanned and the second one on the surface of Melkot who was just constructed atom by atom out of the local atmosphere. Bit of a dilemma. Do you "erase" the original? I don't think the original would like that, I don't care how logical he is.

Ever the physicist, Krauss points out that there is a way to deal with this issue. "Let's just say I vaporize the original at the same time I am creating the copy elsewhere. This way no one feels murdered. Except that vaporizing you produces a huge amount of energy. (You could shoot them or melt them, but that seems a lot like murder.) So you're back to the same environmental problem you had with the nondigital approach. You need the energy of a thousand one megaton nuclear weapons to dispose of the original at exactly the same time you are creating the new version. But then that's not my worry," says Krauss, grinning again. "That's an engineering problem. Luckily, I'm a physicist."

So it figures . . . the one *Star Trek* technology I was really hoping would be right around the bend, turns out to be the most difficult one to pull off. I abhor flying, I abhor traffic, I abhor waiting and wasting time. Here we have the best possible way to get exactly where we want to go with absolutely no hassle, and it turns out to be, well if not absolutely impossible, damned difficult.

But I'm an optimist. I'm holding out hope. Some bright scientist may yet come up with a way to solve all of these problems. Maybe it'll be some great formula that compresses the information in the human body so it can be gathered and beamed in seconds rather than eons. Maybe we'll arrange to beam a computer ahead of us that would gather and assemble the right atoms so we're accurately reconstructed once transported. Maybe some bright bulb somewhere will actually invent a Heisenberg compensator! What do I know? Nothing, except if anyone has any good ideas out there, I wish you would get cracking because time's wasting, and the traffic here in Los Angeles is unbearable!

Interfaces and I do not get along.

—William Shatner

There is no safety in numbers—or anything else.

—James Thurber

6

TECHNO RANT #1

Heeeeeeeeelllllllllpppp!

That's generally my first reaction when digital technology and I come face-to-interface. Well, truthfully, I don't usually scream out loud. It's more like a muffled whimper. But no matter how much I hide my sniveling, I am certain that others around me notice the symptoms of the *digiphobia* that afflicts me. I break out into a cold sweat. My eyes grow big and white. My hands fumble for a book or newspaper, something . . . old-fashioned.

I can handle the simple stuff. I understand, for example, why I can never open the refrigerator door faster than the light will go on inside. And I'm reasonably good with the toaster. But to me all of these fall into the kitchen-appliance–electric-shaver category and don't really qualify as "technical." Where I get my shorts in a bunch is with things that rely on computer chips to operate. Tiny, silicon apparati shot full of zeroes and ones. In short, things digital.

I can't even say I have a love/hate relationship with these contrivances. It's more like fascination/dread. Whatever it is, it's heartfelt. Once I even had the audacity to tell the legendary scientist and father of the hydrogen bomb, Edward Teller, that when it comes to modern technology I am incapacitated; legless and armless. Here was a man who had routinely advised presidents throughout his long career. He ran with some of the greatest minds of the century: Albert Einstein, John von Neumann, Richard Feynman, J. Robert Oppenheimer. And there we were, sitting in his cozy living room in Palo Alto. I am on this man's home turf, for God's sake, and I say to him, "I'm in despair over technology and its place in the world. The way I see it, technology is killing us, not saving us."

He was ninety-two at the time and if he had been a little younger (or his eyesight a little better) he might have risen up and whacked me on the head with the big staff he walks around with to support himself.

Instead Teller, the soul of diplomacy and patience, pointed out to me that the creation of technology has been a central activity of humans for as long as we've been around. Most of us, he counseled, live much more comfortable lives than our cave-dwelling ancestors as a result. True, he went on, it is a double-edged sword and plenty of technology has been used to serve evil ends, but that was really a human problem, wasn't it, not a technological one.

The job of scientists, Teller said, is to ask questions about how the world and the universe work, to get to the bottom of things, and then explain the answers. It is, however, *not* their job to set policy. That, at least in a democracy, is the responsibility of the people. You wouldn't want to leave policy up to the technocrats, would you?

Afterward, when I had left his house and walked into the cool Palo Alto evening, I found I couldn't really argue with Dr. Teller's conclusions. It's true, mindless technology doesn't force itself on us, mindless humans do. But in my own defense I don't think my disappointment in technology is just the result of me being an old fuddy-duddy who can't tolerate change, or adapt to innovation.

A lot of technology, pure and simple, sucks! Back me up on this, and let's call a spade a spade. Most of it is hard to comprehend. You have to read manuals, you have to get over the "learning curve," you have to deal with the "interface." And, still, the controls or icons or whatever make no sense. All over the world, VCRs are flashing: 12:00 . . . 12:00 . . . 12:00, as if to say, "STUPID HUMAN . . . STUPID HUMAN . . . STUPID HUMAN."

I say *we* aren't stupid, *the machines* are, and I don't care how many bytes or bits or MIPS they have. Damnit, I'm mad as hell, and I'm not going to take it anymore!

But, of course, I do. I have no choice. I live in the twenty-first century. I am enveloped in "digitality."

Case in point—the Geopositioning Satellite System (GPS). GPS is a service that you can arrange to include with some rental cars. Using complex satellite technology, its purpose is to help guide you from where you are to where you want to go. Sounds great, but the experience that my writing partner, Chip, and I had with one is a cautionary tale for the next century.

Listen closely and heed well . . .

Ever Lost

When looking destiny in the face, as a book like this requires, you are not doing your job if you don't make your way into that particular part of the planet where the future has already been fulminating for decades known as Silicon Valley. Silicon Valley is not merely a chunk of real estate, but a state of mind, where computer chips, money, and innovation flow like an irresistible river into the sea of the new millennium.

So one Sunday afternoon that's what I prepared to do and it

became an enlightening lesson in digital-human interaction . . . or the lack thereof. I left my home and headed off for beautiful Burbank airport. Once there I lugged my bags a couple of hundred miles or so down its elongated corridors to the gate where my jet was scheduled to depart. But upon my arrival I was told that the jet would be forty-five minutes late. First problem. Why, I think to myself, in a world brimming with mobile communication devices, did I not know this earlier? I mean what happened to information when you need it. Isn't that the sales pitch for the new age?

Okay, fine. I could wait. I'm a reasonably patient man. So I found my way to an empty departure area and pulled out my little cell phone, an obvious *Star Trek* communicator rip-off. This is, I confess, one of the few gadgets that I can actually operate. After all it *is* a telephone. Old habits die hard I suppose.

There are some technologies even I can utilize, like the communicator. I can also operate its first cousin, my cell phone. These days the cell phone is considerably smaller than the first communicator. *(Doug Drexler)*

A mobile phone really is a brilliant piece of technology. To think that only thirty-five years after the first *Star Trek* episode we would actually have millions of people using these contrivances as if it were the most normal thing in the world. Smaller than the palm of my hand, it can connect me from anywhere in the country with anyone else who has a similar gadget, in seconds. It keeps me in touch with my children and my office and my home. Any time I want, from any location, I can call and pester family, friends, even Leonard Nimoy.

However, today the phone doesn't work. I open it and turn it on and get a dial tone, but when I dial a number, nothing. Silence. The magic that somehow sends a signal from my hand into the ether to connect with another phone is not happening. I feel marooned. No amount of mumbling or button jabbing or rattling brings it to life. This happens to me all the time. Why? Have I somehow offended the digital gods? Did James Kirk disappoint a teenage Bill Gates in a particular *Star Trek* episode?

That was problem number two.

Once again undone by technology, I have no choice but to settle down, stow my communicator and do something else, like read. That was fine. Books are a technology that I can comprehend. They don't require batteries or wires. The display is extremely good, even in marginal light. It boots up instantly when I open it and I can go directly to the page I want. No programs have to load, no operating systems need to operate. Ten seconds after I've opened the book, I was thoroughly engrossed in a great story.

Can we please make the digital future as intuitive as this?

After a while Chip arrives so I shut the book down (close it) and we plan how, once we touch down in Silicon Valley, we will intrepidly unpeel the onion layers of the future. Finally, the plane is ready to depart so we schlep on and in less than an hour schlep back off into San Jose airport to meet our next big challenge: getting a rental car.

And this is where the GPS system comes in. Problem number three.

Given my general technophobia, I was frankly a little nervous about using something called a Geopositioning Satellite System. Sure *sounds* digital. Then again, why not try it out? Chip and I are intrepid futurenauts, right? What better way to get around Silicon Valley, cauldron of high technology, than by tapping into state-of-the-art gizmology like this? So now, giddy with the promise of GPSness we make our way out to the white Ford Taurus we have rented, get in, and begin to plot our course.

Inside the car we find a palm-sized gadget anchored to the dash. It is outfitted with a small screen, and a circular touchpad ringed with arrows, and several keys with words like ENTER and LOCATE and CANCEL. We gaze at it for a long time, stupidly, like a couple of guys in a Bud Lite commercial. Silently, it stares back with its single, cyclops eye. Like Odysseus, I'm not certain whether to negotiate with the thing or put its eye out with a hot stick.

I figure I should at least try communicating first. So we turn it on.

Where, it asks with its icons and backlit screen, do we want to go? If only we knew how to tell it! We have an address, but we can't figure out how to enter it. We know we want to go to a place located in the state of California (that eliminates forty-nine other states and a few protectorates in the Pacific Ocean). We also know we wanted to locate a hotel on the Camino Real, but how do we tell it?

We pull out our trusty paper itinerary and search for a way to somehow explain to the machine where we want it to help us go.

I turn on our tape recorder (after all, we want to preserve this encounter for the ages), set my jaw as the intrepid captain of the rental car, and intone, "We are now going to try to operate the GPS system. Our destination? The Crowne Plaza Cabaña."

Chip: Okay, it says here that first we choose our language.

B: Klingon!

C: (Makes guttural Klingon noises.) It's not working.

B: Try English.

C: Okay. Where, exactly, are we going? This thing is going to demand precision.

B: The Crowne Plaza Cabaña at 4290 El Camino Real in Palo Alto, California. Jeez, that's a mouthful.

Computer: (Suddenly talking in a pleasant female voice.) *Resume route to?*

B: It speaks!

C: She sounds pretty good. Think she's free tonight?

B: Easy. We're working here.

C: It wants us to resume our route.

B: Resume! We haven't even started. We need to punch in the destination. Let's see, what do we press? How about this?

Computer: *Select destination.*

C: Select destination, how?

B: It says to press arrow or enter keys. Okay, we're pressing enter. Let's see . . . popular destinations, product, personal address book, customer support . . .

C: Popular destinations. Hit ENTER.

B: Okay. Burbank, Los Angeles, San Diego . . . (clicking) . . . San Jose!

C: San Jose, good. Do you think it knows the way to San Jose?

B: Selecting San Jose. More destinations. Let's see . . . Japanese Friendship, Lick Observatory, NASA, Pebble Beach, Paramount's Great America . . .

C: There you go, the Rosicrucian Egyptian Museum.

B: Perfect! Rosicrucians communicate by telepathy. We wouldn't need to go through any of this there. We could just *think* our questions and they would *think* back the directions. I like that.

C: Oh-oh, we're way past Crown Plaze Cabaña. I guess the Cabaña's not one of the popular destinations around here.

B: Let's go back to . . . how do you go back?

C: Good question. Here, cancel!

B: Careful. It may be a trap. There, okay. Back to San Jose.

C: Is that in the Alpha Quadrant, Captain?

B: (Dirty look.)

C: Right. Cancel.

B: Now it says, "Return to airport." *That* should be easy, we're only a hundred yards away!

C: New screen.

B: #$@@&*!!! I'm doing what I always do in every technical, computerized, digitized situation I ever get myself into! I reach this impasse, I hit a wall, and I cannot go any further and . . . I give up.

C: You're a starship captain, can't you do something?
B: Scotty, get us out of here!
(Long pause, gazing out the windshield. We do not discombobulate.)
C: We are lost!
B: We are lost!
Computer: *Resume destination to?*
B: Oh shut up!

Needless to say we were not off to a good start. Actually we weren't off to *any* kind of start! For ten minutes, sitting carbound a stone's throw from the airport exit, we have been punching buttons without success, feeling more stupid by the second. Here we have this great technology right there in our hands and we can't figure out how to communicate with it! Here I am, a man who has at least pretended to routinely blast through the galaxy at warp speed and *I cannot lock in a course for the Crowne Plaza Cabaña!*

Classic problem. The technology is great. The design sucks.

Deep breath. We still need a place to sleep. But Silicon Valley is big. We can't just bounce around like a pinball and hope we bump into the elusive Crowne Plaza. We need help. So I put the car in drive and, swearing at the cyclops like a couple of Barbary pirates, Chip and I head onto the freeway in search of the only interface we feel capable of comprehending—a human being.

A few miles down the road we're scanning the darkening horizon for carbon-based life-forms. Everywhere lies the evidence that we are in the heart of Silicon Valley, Innovation Central, Geeksville, handmaiden to the Digital Age. Along the flat land that lies on either side of the freeway, rise sleek glass and stone buildings wired, I imagine, to the teeth, silently pumping out zeroes and ones like old rust-belt industries pump out carcinogens and carbon dioxide. Lots of buildings.

People, however, seem to be in short supply.

We pick an exit at random and pull off. A cluster of gleaming high-rises stands like sentinels, silent and empty. We drive down a wide boulevard surrounded by acres of office parks and empty lots. At one point we are startled by a human voice, but, it's a false alarm, not really human, only a PA system at a closed Toyota dealership, eerily blaring out a recorded message in Japanese to vacant squares of asphalt.

At last we approach an intersection. With our GPS system silent and mocking, we spy that ever-reliable twentieth century navigational device—the gas station. Sad to say, we have now been reduced to doing the one thing that no real man ever wants to do: ask for directions.

There were, of course, no attendants at the station. Digital credit-card–operated pumps have disposed of jiffy service. However, one lone soul stands by his compact car pumping it full of gas. I roll down the window.

"Excuse me, sir? Can you tell me where Camino Real is?"

The stranger looks at me. "Yeah, it's . . ."

"By the way," I interrupt, "what's your name?"

Now the guy's getting a little nervous. *He* looks around to see if there are any human beings nearby. Is he going to be jumped or is this *Candid Camera?* "My name is Hyojin," he says.

"Hyojin, I've got a GPS system here. Guarantees you can't get lost in a million years, yet . . . I can't find blankety-blank El Camino Real. Can you help? But first, because I'm so thrilled to find a human being, tell me what you do."

"Real estate sales."

"You're a real estate salesman. You have contact with human beings then?"

"Yes."

"You don't deal with machinery?"

"I have computers. I deal with computers."

"Do you like computers?"

"Yeah."

"But you don't use them in your real estate sales. It's one on one."

"Right."

"You say, 'Buy this property, it's really valuable. It's going up in price. You'll make a fortune.' "

"But then we end up creating more business and more housing and more people and more traffic," says Hyojin, pumping away.

"Which is why, I suppose, we need clever little technologies like a GPS."

Smiling. "I guess."

"So we can *never* escape the computer. Yet now I've come begging to this human, you, to help me, because this thing—I don't, I don't . . . So now, human to human, which way do I go to get to El Camino Real?"

He looks us in the eye and starts to give us directions.

I like this Hyojin interface. He knows how to get us to where we want to go. He doesn't require any button pushing. No inscrutable icons. He smiles and immediately understands my questions. When I speak, he listens, considers what I'm saying, and answers me clearly within a few seconds. He is very effective at getting the information from *his* database into *my* database, and once he does, we are able to get a bead on our hotel.

Heartily, one human to another, we thank him and set off for the Cabaña, but not before stopping at a sushi bar for dinner. By this time I was dying to sit down in a restaurant where I know that each serving will be lovingly, humanly prepared by hand right before our very eyes. This, I thought, is the antithesis of the faceless technology that has been plaguing us all day. This food will be made to our individual specifications. It will be personal. Or so we thought.

Instead when we walked through the doors of the place we discovered we had arrived at one of those new sushi restaurants; the kind outfitted with tiny rivers that flow around the counters like mini-Mississippis with barges of sushi floating by like so many junks in a Hong Kong harbor. Rather than watching someone individually prepare our meal, we were instead presented with this buoyant buffet from which we dejectedly chose our facelessly prepared sashimi. No conversation with the chefs, no eye contact, not a snippet of interpersonal communication.

Dejectedly we gazed at our boat of food, and ate.

It was clearly just going to be one of those twenty-first century, done-in-by-technology days. No way around it. So we filled our bellies and headed to our hotel, following Hyojin's impeccable directions. Once there we wearily checked in, unpacked our bags, and, at last, went to sleep, dreaming of what the future may hold.

Of course, before I got in the room, I did struggle briefly with the damned electronic key!

PART TWO

THE BITSTORM

The best way to predict the future is to invent it.

—Alan Kay

In which we discover that ordinary things may soon become extraordinarily smart, computing may become a part of us, and we a part of it. Will interfaces (and Windows) finally bite the dust? Is resistance futile after all?

Today nature has slipped, perhaps finally,
beyond our field of vision.

—O. B. Hardison Jr.

7

BITSTORM

The Image:

Kirk, in crisis. He stands on the bridge of the *Enterprise*, *determined*, jaw squarely set, barking orders—warp drive this, send-a-message-to-Starfleet-that. Throughout this climax, all 190,000 tons of the *Enterprise* operate exquisitely and hurtle along faster than the speed of light. The artificial gravity system works, every person and communicator and computer is linked flawlessly and invisibly. Deflector shields guarantee cosmic debris or enemy attacks don't transform us into an interstellar scrap heap.

How was all of this possible way back in the 1960s when transistor radios were still considered the height of high technology? Well, it wasn't.

The Reality.

Sure the deck consoles are humming away, and the impression is that a matrix of highly advanced technology is working as smoothly as a twenty-third century Rolex. But the ugly, unvarnished truth is that the high technology driving the *Enterprise*'s rhythmically blinking lights was really some guy named Joe or Bob or Elmer hunkered down under the console running a nail attached to a wire back and forth along each of the connections like a kid running a stick along a picket fence. (Look closely at the lights sometime and you'll notice they aren't perfectly in sync with the sound effects.)

Back in the early days of the series even air conditioning was considered high technology. I remember one day Bob Justman, the show's tireless co-producer, came down to the set because a garbled message gave him the impression that Shatner and Nimoy had said it was too hot to shoot. "The heat," he was told. "They just get too hot, and they won't work."

Justman bolted out of his office and rushed up to Stage B. Prima donnas, he fumed as he stomped to the set. I'll give them hot. Damnit, this is a television show, not a circus. Money is getting burned by the armload every minute we're not shooting.

By the time he arrived on the set, he was loaded for bear. He walks up to Leonard and me and says, "What's the matter, guys? You have a problem?"

Well, we told him, *we* don't have a problem but *you* do. Your equipment is overheating.

"My equipment?" says Bob.

It wasn't us, it was the electronics that powered the lights and readouts on the bridge set that were overheating. In 1966 there were no integrated circuits, and certainly no silicon chips, at least not on the Desilu lot. Everything was run by old vacuum tubes like the ones that operated televisions in those days. Remember going down to the hardware store with dad to test the television tubes to see which ones had gone bad? Same kind of tubes, and these suckers got hot, especially crammed in under the fake consoles. And when they got *too* hot (which didn't take long), they blew out and any scene that was being shot ended right then and there.

So what did the ever resourceful Bob Justman do? He brought in a battery of air conditioners to keep the consoles cool, turned them off when we were recording a take (they were too noisy), and then we all prayed like hell that the tubes would stand up until the take was complete.

That was the *real* state of the technological arts in 1966.

So maybe I should rephrase the original question. How was all of this high onboard technology *supposed* to be possible? Before starting work on this book, I must tell you that I couldn't even guess. And I certainly didn't know in my early days playing Captain Kirk. Back then I was just trying to hit my marks and not blow my lines. But today, after talking with all of the people that I have talked with, and after seeing all that I have seen, I know the one-word answer.

Computers.

The whole warp-factoring ship was thrumming with hypothetical computers. Millions of them, all sizes and shapes, from powerful mainframes to teensy-weensy sensors, everywhere, invisibly communicating, and crunching numbers at prodigious rates. The imagined *Enterprise* bristled with digits flying between man and machine, machine and machine, machine and the rest of the universe.

In this world everything was digitally enhanced and thoroughly linked. We had rooms that knew when we arrived, communication devices that let us commune at distances anywhere from a few feet to several light-years, gadgets that sensed virtually everything, computers that talked and robots, androids, and cyborgs conjured from silicon, circuits, and the most exotic possible alloys.

None of us was ever very far from some digital thingamabob. Bones had his scanners and computerized sickbay. I routinely signed off on orders using a digital notepad (later known as the padd) handed to me by my trusty yeoman. Scotty's consoles ran the warp engines and beamed scattered bits of us across space and then reassembled them (usually) at their appointed destinations in the proper combination. Our communicators tracked and located us and allowed us to instantly transmit plot sensitive information from planet or shuttle or ship deck. We even had personal computers in our own quarters (and never once, I swear, did I visit *Playboy*'s World Wide Web site . . . except to read the articles). If megabytes had been shrapnel on the *Enterprise*, we'd have all been cut to pieces.

Yet notice that in *Star Trek*'s imagined future most computing was invisible. Yes, we were digitally enveloped, part of a vast storm of information flowing in waves throughout the ship, but the computing was not obvious; it was everywhere, but it was nowhere to be seen.

Once again, *Star Trek* seems to have hit the futuristic nail on the head because this invisible, digitally enveloped world is exactly where we seem to be headed today.

Smaller Is Better

Only recently have I begun to notice, at least consciously, something that has probably been blatantly obvious to everyone else for some time. Since computers first entered the picture more than a half century ago, they have been getting not larger, like commercial jets and SUVs, but smaller. They are growing less monolithic, more dispersed, increasingly connected. I walk through airports or sit in restaurants and people are talking on their cell phones or checking pagers or wirelessly communing with the Internet using some personal digital assistant. (As you read this nearly 10 million people are accessing the Internet with a handheld gadget. Two years ago almost nobody was.) Computers are in our cars, our televisions, and our microwaves. We carry them and pocket them, all while they incessantly chatter away at one another, or they allow *us* to chatter with one another regardless of where we are. When I see this, I can't help being reminded of Kirk, Spock, McCoy, and the rest of the *Enterprise*, all digitally affiliated. Thirty-five years ago this was wild stuff—a future embroidered with zeros and ones. Today your average teenager would think it odd that I even notice any of it. Isn't everyone connected? Haven't people always and instantaneously been able to communicate?

Well, no.

But it sure seems to be where we are headed. It's this *Star Trek* style of intercommunication between humans and machines, and machines and machines that is leading us into what seems to me to be an invisible blizzard of information, a bitstorm, for want of a better term. It feels strangely familiar.

How do *I* know all of this? Well, I don't *know* anything! I'm just telling you what I've learned. And I am also telling you that all of this "digitality" doesn't necessarily warm the cockles (whatever *those* are) of my technologically challenged heart. Remember, I am a high order techno-greenhorn, and to be blunt I find this much computing damned disconcerting. It was fine when it was all smoke and mirrors, but I'm not too sure how I feel about it permeating the *real* world. How far will it go? Can a future this digital be benign? Will it mean we'll have more time to do what we want to do, while our digital servants handle the minutiae? Or will it mean we are simply walking blithely, gadgets in hand, into *The Matrix* where our humanity will be obliterated in a storm of bits? On the other hand I'm not sure that what *I* think really matters. The brewing bitstorm seems inevitable, like a force of nature.

In the face of that, all we can do is keep searching for answers, keep exploring like the old *Enterprise* crew itself. And that's why Chip and I had journeyed, thinking caps in hand, into Silicon Valley. We wanted to visit one of the meccas of innovation, Xerox PARC (Palo Alto Research Center), a place where the future bubbles continuously, like a fine New Orleans gumbo.

Any sufficiently advanced technology is indistinguishable from magic.

—Arthur C. Clarke

8

GET SMART

Xerox PARC is not only located in the heart of Silicon Valley, it is arguably one of the primary reasons Silicon Valley exists in the first place. It came into the world in 1970 when Xerox Corporation, riding high on the enormous popularity of its copying machines, gathered together a team of world-class researchers and gave them the mysterious mission of creating "the architecture of information."

To me the height of information architecture is the yellow pages, but the gurus at Xerox had something deeper in mind.

In *Star Trek* we were never far from some digital thingamabob. A few of the gadgets they're toying with at PARC remind me of what they now refer to as the padd (Personal Access Display Device) on *Star Trek*, except PARC's devices are even more creative. *(Doug Drexler)*

They wanted to take a radical new look at how we invent and communicate information. PARC was the result. At the time, since the vast majority of information still resided on paper, this seemed to be right up Xerox's alley. After all, who knew paper better than the company whose name had literally come to mean "paper copy"?

The basic concept of research centers is pretty fascinating if you stop to think about it. They are basically institutionalized sandboxes where thoughtful, bright people are provided lots of buckets and shovels and other toys and then told to go amuse themselves. The hope, of course, is that in the process of playing they'll invent something that will change the world. Seems like good work if you can get it.

But from the beginning Xerox PARC was not like a lot of other scientific research centers, most of which, especially in the 1970s, concentrated primarily on engineering and hard science, things like chemistry and physics. This is, of course, what most researchers and scientists are good at so why bother with messy "human factors." Too fuzzy. Instead, brilliant minds focused on a technical problem and hammered out sublime engineering solutions after years of exploration, experimentation, and testing. Change then ensued.

Breakthroughs like transistors, silicon chips, lasers, and gene splicing all resulted from industrial strength scientific work done at places like Bell Laboratories, the National Institutes of Health, the IBM Thomas J. Watson Research Center, and major universities funded by government or corporate research grants.

There's nothing wrong with a heavily geek-oriented approach to the future, I suppose, but ultimately the technologies and the gadgets have to work for the rest of us. Technophobe that I am, that's what I like about the people at PARC. They seem to hate the idea of complex interfaces and learning curves almost as much as I do. They think like Gene Roddenberry: Technology is cool, but it's not the main event. It's just a tool designed to make life easier. How refreshing. This is why PARC is not only staffed with world-class computer scientists and engineers, but psychologists, linguists, writers, social scientists, and artists; every one of them thinking outside-the-box thoughts, constantly, in an effort to humanize the future.

For that you gotta love 'em. At least I do.

So far the results have been pretty impressive. For the past thirty years the collective genius at PARC has given us the first personal computer, laser printing, Ethernet, the mouse, the Graphic User Interface, object-oriented programming, and many of the basic Internet protocols used around the world today. It's not an exaggeration to say these inventions have revolutionized the way we all work and play.

Xerox, the corporation, however, is an odd handmaiden to the creation of such revolutionary digital technologies. After all, it's the "Document Company," right? This is a corporation that has generated billions of dollars primarily from copying and printing plain, old-fashioned mashed up wood pulp.

That focus has turned out to be both a blessing and a curse for PARC. In the seventies and eighties, when PARC was generating highly innovative computer technologies, Xerox's corporate headquarters had a hard time understanding how anything digital could possibly help them sell more machines that copied paper. I mean, the very concept of a "paperless office" must have struck terror into the hearts of its legendary sales force. Wildly creative technologies like Ethernet cables and software that linked computers into networks that could in turn allow people to share their work *without paper?* What was that all about? Are these guys crazy? Laser printing? Yes, good. Stay on that one. Printing creates more paper and it makes the copiers better and faster, but that digital stuff . . . forget it!

On the other hand . . .

On the other hand, when it came to ruminating on things digital, Xerox's paper tradition gave PARC's thinkers a different, more people-friendly view of the world than your average computer-oriented propeller head. Paper, they saw, was one of the great inventions in human history. It's portable and flexible. It requires no plugs or cords or backlit screens or batteries. You can draw on it, write on it, print words, pictures, or blueprints on it, then fold it up and carry it around for use later on. And it's simple; absolutely intuitive. Never once will a piece of paper "crash" on you. It is so easy to use that you don't even think about it. Paper's learning curve is zilch, nada, goose egg. Sure, maybe six thousand years ago, when paper and writing were first being invented, the papyrus used to "crash." I can imagine Egyptian scribes moaning now. "Man, my scroll went down *again!*" And maybe your average Sumerian couldn't quite get the hang of representing thoughts with squiggles of black dye. "Hey, does that mean sarcophagus or sacrifice?" But for the most part, these days anyhow, the learning curve for using paper is pretty flat.

There's another thing about paper (and words) that you may have noticed. They are so cheap and useful that both are absolutely everywhere. You don't go into your office and boot up *the* paper. After all, the supply is unlimited. You can have all you want, when you want. We can thank a goldsmith named Johannes Gutenberg for this ubiquity. By combining movable type *and* paper he made it possible to distribute knowledge and information all over the place, from the Bible to the wrappers on candy bars. This is why paper keeps outflanking efforts to create a paperless world. It's just too good at what it does. At PARC, they call these various attributes the "affordances" of paper.

But, and I say this begrudgingly, digital technologies have some good points too as I learned from the researchers at

PARC. Once something is in digital form, it's easier to duplicate than a tribble. It can be moved, at light speed, and with supreme accuracy from one place to another—across an office or across a planet, even *between* planets. There's a magical quality to it.

Digital information is also much more malleable than information in the paper world. Any data that is digital—words, video, images—can be combined and recombined, programmed and linked in countless ways. You can slap still images in video, video within print, audio over still images. You can communicate from one machine to another, even from machine to phone to brain. It all links up like Duplos. Finally, because they can be changed, moved, and copied so easily, zeroes and ones are also very efficient and cheap. In other words, digital technologies have "affordances" too.

Given this odd marriage of paper and digits, given all of the questions about how information lives and resides in the world, PARC's researchers began to imagine a future where computers would be so small and so interconnected that the line between the digital and physical worlds would one day go *pffft!* Is it possible, they asked, to embed digital technologies as seamlessly and simply into the real world as we have already embedded paper and print . . . but add digital speed and interactivity to it as well? In short, they asked, what happens to the world if you combine the elegant affordances of paper with the many advantages of digital technology?

The answer? Every object and piece of information would start to get smart.

Star Trek and PARC have obviously envisioned similar futures; ones in which everything is invisibly bathed in digits, but not in a frightening cyberpunk sort of way. Both imagine a world where computation is everywhere, yet, like paper and print, nowhere. A place where the worlds of the digit and the molecule blend so seamlessly that you can't tell where one ends and the other begins.

The implications of this are pretty wild and I'm happy to report that the future envisioned at PARC doesn't look anything like Windows 2000.

Ubiquitous Computing

In the late 1980s and early nineties a PARC researcher named Mark Weiser started pushing the idea of computers being embroidered into the environment in countless physical objects all around us. His hope was that he could extend the "architecture of information" concept that was the bedrock purpose of PARC's existence to include more than paper, or more precisely to weave the information on paper, and everywhere else, right into the environment we live in.

He kicked his concepts off in a 1991 *Scientific American* arti-

Sure, laugh, but I'm telling you, we're beaming up! From left to right, some of the great folks Chip and I met when visiting PARC: Johan de Kleer, Anne Balsamo, Lois Wong, me, Eric Saund, and John Gilbert.

cle. "The most profound technologies," he wrote, "are those that disappear. They weave themselves into the fabric of everyday life until they are indistinguishable from it."

Given how small and cheap and powerful computing is becoming, why, he wanted to know, should computers exist only in the forms of desktops or televisions or even cell phones? The goal shouldn't be to simply take increasingly smaller computers to the beach, jungle, or airport (pretty much what we are already doing right now). The real goal should be to slip computing entirely out of view; make it blend into the background. We should never have to get into the computer's world. Instead computers should disappear entirely into ours, like the birds and the bees.

No more interfaces! Hallelujah!

Weiser called his idea "ubiquitous computing." Maybe not a term that Madison Avenue advertising pundits would embrace, but it was descriptive. Mark provided a peek into what such a thoroughly digital world would be like in the same *Scientific American* article. When I read it I was struck by how Roddenberry-esque his vision was. In Weiser's future, the technologies—the zap guns as Gene would have called them—drew no attention at all to themselves. Instead they were simply there, doing their job and otherwise staying out of your face.

When a fictitious woman (we'll call her Martha) in Weiser's futuristic scenario is reading the paper, for example, she spots an interesting quote. Does she pull out a computer or a palmtop and write the quote into it? No. Does she whip out her cell phone and make an audio note? No. She simply circles it, and the pen sends a message to the paper that in turn transmits it to her office where it awaits her. How can this be? Because the paper is digital and wirelessly linked to a digital network which is, in turn, linked to her office. Thus a simple gesture communicates information at the speed of light.

This kind of digital technology I can relate to. No keyboard. No "on" button. I don't even have to plug anything in.

Tragically Mark died in 1999 of cancer and all of PARC took it hard. But his colleagues continue to extend his vision of a digital world. Roy Want—who currently is a principal engineer at Intel Research—worked as a principal scientist at PARC. He and his team have conceived of several technologies that Kirk and Picard, Sisko and Janeway would be very comfortable with.

Augment Me

Roy told us that his goal, like Mark's, is "to link the physical world with information in the computational world." He calls this "augmenting reality." Do I know what any of this means? Uh-uh, but I'm sitting in Roy's office trying mightily to look intelligent as he explains it to me with great patience and a warm smile.

One of the things about the researchers at PARC—they don't just talk. They build real gadgets and gizmos. And they stick them out there in the world and see what they do. If they work, they refine them. If they don't, they trash them and go back to the drawing board. It's a variation on the laws of evolution: Survival of the fittest . . . technology.

Earlier in the nineties, for example, when Want was on Weiser's team, they built and placed infrared sensors throughout PARC's offices and then distributed electronic badges, *Star Trek* style, for employees. The badges "tagged" or identified them when they walked into certain rooms around the research center. Sensors in the room then "understood" who was there and set the lights and heat to their liking. It turned out the technology worked pretty well, but the employees didn't like being ID'd. It smacked too much of Big Brother. So the program was ended.

But they soon found different uses for the badges. Instead of attaching them to people, they started attaching them to things and turned them into something called PHICONs.

What, you ask, is a PHICON?

Fair question, so I put it to Roy. Not an alien from the Delta Quadrant, I was happy to learn. "It's a PH(ysical) ICON," says Roy. "You're familiar with desktop computer icons, right?"

"Barely."

They're symbols, Roy patiently explains, that represent various kinds of documents or programs on your computer desktop. Xerox came up with this idea back in the seventies and called it the Graphic User Interface (GUI). Anyhow, says Roy, PHICONS simply extend the concept beyond the desktop. Instead of moving symbolic representations around on a computer screen with a mouse, you move them around in the real world. He gave me an example.

Say you have one of these little PHICONS in your pocket and you go into a meeting. In the room is a large electronic whiteboard. This board is essentially a big computer screen that is linked into a network. Let's further say the PHICON in your pocket represents a document you've been working on. Now you want to share the document with everyone in the meeting because *it is a work of pure genius!* You pull the little PHICON out, wave it at the white board, and *bam!* there is your masterpiece for everyone to see. The sensor on the whiteboard recognizes the document from the PHICON, finds it on the network, and zaps it onto the board. It's as though you had reached into your computer, grabbed the icon that represents the document, and put it in your pocket. And then like magic you could just toss it right up there on the whiteboard.

Now, because everyone is dying to read the paper, they can wave any PHICON they happen to be carrying at the board, and now that PHICON knows all about your document. All your fellow workers have to do now is wave the same PHICON at their own computer (home or office), and *Presto!* your magnum opus pops up right on their screen. Amazing.

Unfortunately not everyone agrees that your document is a work of unparalleled genius and they tear it to shreds. You become despondent, take to gambling, liquor, and listening to bad country and western music, and . . . But that's another story.

The point here, says Roy, is that we already have all the interfaces we need in the real world so why muck them up with more on computer screens and VCRs if we don't have to? In the real world if we want to mark a book, we use a bookmark. If you want to leave a note for someone, you grab a scrap of paper, write it, and leave it on the kitchen table. Now what if you take these physical things and also make them digital? After all, a big advantage of digital information is that it can be connected. Paper is easy to use, but dumb. Literally mute. It can't communicate except through the relatively cumbersome interface of our own minds and memories, which we all know are flawed, and frankly have better things to do than track minutiae scribbled on shreds of paper. So, asks Roy, what if everything—paper, words, images, machines, and computers—could all be linked and accessed with very little effort?

I asked Roy if he had a real-world example, something I could get my pea brain around. Sure, he says. Imagine you're working in your office, you know, the corner one on the fiftieth floor with the spectacular view. You have a desktop computer and a handheld PDA (palmtop)—something, perhaps, that looks similar to the clipboard-like padd (Personal Access Display Device) that is used so often on the bridge of the *Enterprise*.

This padd holds everything from your appointments to documents you happen to be working on. In your office of course there are also paper versions of everything strewn all around you (remember this is a Xerox scenario). Besides most offices, despite all of their high-powered, networked desktop computers, are still knee-deep in paper.

Anyhow, you are summoned to a meeting where you'll be discussing the *Parallel Universe Travel Project* that your crack technology team has been working on. Before you go, though, you want to find the latest version of the *Wormholes for Fun and Profit* white paper you've been writing that describes in scintillating detail all aspects of the project. You can't find the latest version, but you do have an old paper version on your desk. Great. You simply pick it up and pass it over your padd or PDA or whatever, like a magic wand and, voilà, the latest version appears on the screen and off you go to the meeting.

Along the way, you run into a coworker who shows you a book she's reading: *I'm Working on That!* Son of a gun! This is the greatest book since *Get a Life*, she says. You have to buy it. Since she won't lend you her copy (she probably realizes I need the royalties), you skim through it, and agree life would be meaningless unless you can get a copy lickety-split. You wave the book over your PDA and *poof!* the book is instantly ordered over the Internet (assuming you beat all of those mail fraud charges and you're credit rating has been restored). You can buy either the physical version of the book or have a digital version delivered right to your PDA. (The future is filled with options.) And notice that even though much of what you are doing is digital, you have yet to touch a keyboard or leap a single learning curve.

But there's more!

Farther down the hall you see a poster for a *Star Trek* convention. Now how can you pass *that* up? Naturally you want more information so you point the PDA at the convention's title and it immediately loads the convention's Web site right there on the screen you have in hand. You pull the paper bookmark out, tap the PDA with it, and instantly the site is bookmarked so you can check it out later when you have more time.

Now you point the PDA at the poster where the convention's date is printed. Instantly it loads all the information into your calendar, including where the convention is located and how long it will take to get there. It'll even generate a map. (Or you can have the GPS system guide you there.) Again no inputs, no buttons, no interfaces.

How is all of this magic possible? Roy holds the answer up between his thumb and index finger: just a little glass capsule Roy calls an e-tag. It's hardly larger than a staple. It takes no batteries because it's powered by the signal used to access it. When it receives a signal from a printer or a computer or a PDA, it turns on, like an electronic firefly, and then having sent its own signal it turns off again.

In Want's future, the world would be loaded with countless e-tags like this one. Eventually, he says these little fellows will grow even smaller and become embedded in the standard production and printing processes for all sorts of objects and appliances. It'll then become possible to weave computing power into more paper, documents, and posters than you can shake a mouse at. Once that's done they can all be linked to PDAs, televisions, cell phones, earrings, watches—any digital gadget we create. Everything will start coming alive and connecting, tossing bit streams back and forth like lasers at a light show. In this world the bitstorm would be invisible, but very heavy *and* very *Trek*-like.

I just stare at the tiny thing. It's at moments like this that I just can't resist saying, "What *will* they think of next?"

You would think that this digitally interwoven future would be good enough for the folks at PARC? But no, they have to be innovating all the time. New this, new that. The place is crawling with people who have irrepressible, chronic cases of *What-if-itis*. What if we start not simply digitizing and connecting paper and gadgets with tiny e-tags, but just pretty much start connecting everything? What if we go beyond documents and start to mix the invisible power of computing into the walls, furniture, floors—the objects and surfaces all around us? What would that be like?

Okay. I'm game. What?

Well, for that we have to go talk with Andy Berlin.

The IQ of Things

Andy Berlin looks about fifteen, bright-eyed and brimming with ideas. He's actually in his thirties and was a senior researcher at Xerox PARC when I spoke to him; now he's a principal engineer researching Smart Matter at Intel Research.

Berlin has a twinkle in his eyes because the world around him looks like a playground; one where he can turn everything into a toy of sorts. Andy dreams of raising the IQ of things. All kinds of things. He wants to make matter smart.

So, he wonders, what if everyday objects acted like they were alive? In such a world you could move things around on your desk just by talking to them or directing them with your hands, like Zubin Mehta directing the New York Philharmonic. Cracks in walls would repair themselves like a cut on your thumb heals. If you walked into your office and were feeling a little blue, you might command the walls to brighten up, or change their color entirely, or even display a Hawaiian sunset. Paintings and pictures at home could be changed instantly, drawing from a library of images you have on file. And we aren't talking about video displays of pictures either; the walls (or rather the paint on them) would actually, physically change! Or what if the wing of an airliner were bending under stress and in danger of cracking? In Andy Berlin's future, it would flex itself in the opposite direction, all on its own, like a living thing because the material in the wing would be computerized and smart enough to know how to make itself stronger and safer.

Essentially all of the stuff that we normally think of as just passive chunks of material—plain old objects—would become "programmable," as Andy puts it. In a word, computerized, except there wouldn't be anything in sight that you or I would actually call a computer.

When you look at Andy, he doesn't really appear to be mad. But after I heard him spouting these things I had to wonder if maybe he had been spending a little too much time in front of the old work station or maybe sniffing the toner down the hall by the color copying machine. (At Xerox there's a lot of toner.) But, in truth, there actually is a technology that is being developed that could make exactly these kinds of things (and much more) possible within five to ten years. It's called MEMS, Microelectromechanical Systems.

MEMS essentially employ what Andy calls micromanufacturing techniques, like those used to cram thousands of transistors onto outrageously small computer chips. Except in this case the technique is used to make simple, but very small parts that bend or turn or change color or "smell and see" depending on what sort of electrical current they receive or chemicals they come into contact with.

Basically, says Andy, adjusting his glasses and looking at me as if I understand what he's talking about, MEMS technologies combine the capacities of very small computer chips with these extremely small machines and then link them electronically as well as physically. Researchers have already made working mirrors and lasers and airfoils small enough to fit on your fingernail!

I couldn't get my mind around this idea really. How can objects so small that they can be dwarfed by a single dandelion seed interact and communicate with one another? But Andy explained that there is no physical reason why machines and sensors and actuators can't work at this scale. We just have to make the working parts very small. Computer chips work just fine even though they are microscopic in size. Machines will too. You could make smart dust, for example—tiny speckles of metal loaded with microscopic sensors that might communicate with one another if they find a lethal gas in the atmosphere. The military is exploring this as a defense against biological weapons. The same sort of thing could be used to sense water on Mars, or the internal workings of a tornado.

Or you could create a car bumper made of millions and millions of tiny metal machines, each of which knows its proper shape in relation to the next one. If you smash the bumper when you back into a utility pole at the shopping center, instead of calling your insurance adjuster, you just watch as the bumper "remembers" the way it's supposed to look and morphs back into its original shape. (I wish my body would do this and remember what it looked like when it was twenty-five.) It would take a small amount of electrical power to give the machines the juice they need to reconfigure themselves because each of the individual pieces are so small.

Then there's my favorite scenario that Andy spun out for me. In the not very distant future (certainly sooner than the twenty-third century), imagine that you are in your office. Recently you have had the walls covered with a special "paint" that transforms them into one big interactive display. How do you manage that? Essentially, Andy explains, you cover the walls with millions of laminated MEMs that can display images as well as sense movement and receive electronic signals. The tiny displays consist of encapsulated chemicals a couple of microns square, each one roughly the size of a human cell. This means that if you lined up 25,000 of these tiny gadgets, they would add up to one inch. When you feed each of them a different level of electrical current, says Andy, they turn different colors and that turns your wall into a full-color, super-high resolution display—instant interactive wallpaper!

So there you sit surrounded by your digital walls. With a word, you boot them up. Drawings, notes, and documents that you've been working on appear all around you. Since the walls sense you as much as you sense them, you can move whatever is represented on the wall around, with a gesture or with your computer mouse (if you're still using one) or even directly with your hands by simply walking up to the wall and pushing things this way and that. (For this you may have invisibly small, micron-sized gadgets called accelerometers painted on your fingernails; tiny computers that interact with the wall's sensors and tell it where your hands are and what they are doing.) Just like the days when you were a kid with a crayon, you can write on the walls and connect not only the dots but every idea that your little heart desires.

But it doesn't stop there, says Andy with glee. The walls, thoroughly computerized by their special "paint," are also linked to your computer network which holds the originals of all of your digitally embodied thinking. As you change your drawings, documents, and comments around on the wall, each is also changed and updated in the network. This then makes it possible to easily send part of your wall (and mind) to someone else's wall (and mind) via the Internet. It even makes it possible to collaborate with people anywhere in the world, in real time, without ever moving from your chair. They'll see you, you'll see them; it'll almost be like you're both in the same place. And in a way you are because the digital worlds inside the network, your mind, and the physical space around you have all been seamlessly married.

"Which is exactly what this is all about—joining the physical and digital worlds," say Andy.

Sound far-out? Does to me. Sounds impossible, in fact. I was a kid during World War II. In those days resources were so scarce paper *was* valuable! Radio was the height of technology and the closest my walls came to displaying images was when I shadow boxed with the self-projected image of my body made possible at night from the streetlight outside my window. So the idea of objects and walls coming alive with sound and images, of objects morphing and reshaping right before your eyes, well that seems a little spooky to me.

However, I was assured that interactive walls are not as far-out as I think. Then I was provided the proof when I was introduced to PARC's Nick Sheridon.

Electronic Paper

Nick Sheridon and his research team have invented a working version of something called Gyricon, also known as electronic reusable paper. It is basically paper that acts like a computer, and the thing about it is that it's not a pipe dream or even an experiment. It works right now.

Gyricon isn't much thicker than your average piece of typing paper, but it's made out of something entirely different, two very

Gyricon is a form of electronic ink. Each thin "sheet" of digital paper is loaded with millions of beads that are black on one side and white on the other. When they are provided with an electrical charge, they flip into the right position to create whatever document you like.

thin sheets of sealed plastic embedded with millions of tiny polyethylene balls. Nick places a sheet in front of me and asks me to look closely. Each of these little balls is white on one side and black on the other (or they can be red and white or some other contrasting color) and they swim in a thin bath of oil so that they can easily flip over, revealing either their light or dark sides.

"Each of the beads is electrically charged, which means if you send one a signal to flip over to its black side, it will, or vice versa," explains Nick. "By orchestrating the beads in various patterns you can create a document or a picture, just the way a computer screen or a newspaper uses millions of black dots (and white spaces) to represent the type or images that you see when you look at them. Where you want letters and lines, the dots flip over to their black side. Where you want white space, they flip over to their white side."

Nick Sheridon unrolls his electronic paper. Now his invention has found its way into the real world as Gyricon Media, Inc.

I finger the plastic. It's very slim. The little dots just sit in there at random, dumb as dominos waiting to be flipped. "But how do you 'write' on the paper. How does it know what to display?"

Nick explains that for now he sees using a kind of magic wand to make this electronic paper work. Wand seems like an appropriate name.

He holds up a prototype. It's plastic, about one inch thick and half a foot long. Unlike other magic wands, this one needs batteries. Nevertheless it still has properties that would impress Harry Potter. In this little stick you can store information for up to ten books! You harvest the books by downloading the manuscripts wirelessly from your computer. Just set the wand by the computer and let it commune with an online bookstore or newsstand. Once the wand is "loaded" you simply swipe it over the paper. An electrical signal is passed along to the tiny oil-encapsulated balls inside the electronic paper and each flips over into the right configuration to deliver a perfectly printed page that requires no power, no back lighting, no integrated circuits. This takes portable computing to a whole new level, and not once do I need to hit CONTROL-DELETE!

So I'm thinking, I go on a trip. The transporter hasn't been invented yet so I know I'll have some time on my hands. Rather than loading my bags down with backbreaking, brick tomes, I fold a piece of electronic paper or two up into my pocket and toss my "word wand" into my fanny pack along with my mobile phone. (Forget the PDA. Now I don't need it.) Once on the plane, I swipe the wand over the paper and it "prints" the first page of the book. When I'm finished, I print the next one. The same could apply to the day's newspaper or an issue of *People* magazine, or, for that matter, your favorite web site. The beauty of this Gyricon stuff is that it requires no batteries; it's completely portable and unbreakable; I can roll it up like a magazine, yet load it with all sorts of information, tailored just to my needs.

Soon Nick envisions doing away with the wand completely. Eventually, as costs drop, he hopes you will be able to download any information you want directly into the paper itself. When you want to go to the next page, rather than swiping the wand, just tap a corner of the "paper" with your finger as though you were turning it, and the next page appears.

I could get used to this.

Imagine sitting down every morning with a cup of coffee and several big sheets of electronic paper which have already downloaded today's news. In a hurry? Just fold the paper up and take it with you. Here in one slim bag of charged beads you've combined the elegant affordances of paper, print, and computers. You've melded the digital and molecular worlds.

This stuff is already pretty cheap to manufacture, no more than a couple of dollars for a sheet the size of a standard piece of typing paper. (The costs will only come down and you can reuse Gyricon thousands of times before it has to be tossed.) It is also very durable so it won't be long before you can apply it to the walls of your office, like wallpaper where it can display all of your great thoughts.

Being an avowed technophobe, I actually get a little giddy when I see this kind of thinking because it doesn't feel so damned "electronic." It shows that computing can be about more than Web sites, laptops, and cell phones. Now this early version of electronic paper is pretty clunky, just as early fax machines and printers and cell phones were. It's not going to let you recreate Andy Berlin's sunset on the beach, but radio started out with crystal sets and television started out with rabbit ears and little black-and-white screens. So I can easily imagine that it won't be long before we're all using this stuff (or something very much like it) to get our daily doses of information. I'll take two.

Living Things

After talking with Roy and Andy and Nick, I thought to myself: this stuff out *Star Treks* even *Star Trek*! I can't think of anything in the series that quite anticipated innovations like these, and this is just the tip of the iceberg. Chip and I talked about it all on the way back to the hotel. The way we figured it, if matter becomes as smart as these researchers foresee, our environments will grow magical. It would be as though we are living in a fairy tale where objects all around us are alive, as aware of us as we are of them.

What I love about this approach is that it's so . . . simple. Everything is intuitive, seamlessly replacing the frustrating interfaces of computing with the intuitive ways in which we automatically move though the "real" world. Not only does this eliminate DOS prompts, it doesn't even require that I know how to operate a mouse. Even I, who has to be carefully schooled to understand how to turn my laptop on, can imagine how a smart world might work.

If this Xerox PARC/*Star Trek* future actually unfolds I can see us evolving a kind of digital ecosystem that may soon rival and then outstrip the complexity of the "natural" one that made us possible in the first place. In a weird way we are creating a menagerie of digital creatures, species of all shapes and sizes, each increasingly connected and reliant on one another, with ourselves right in the middle of it all.

Is this progress? Got me, but I know this: there's a lot more to come.

More?

Yes. But to experience it you have to leave Silicon Valley and head off for the other end of the continent.

I must go down to the sea again,
To the lonely sea and sky,
And all I ask is a tall ship,
And a star to steer her by.

—John Masefield

Where do we make the left turn again?

—William Shatner
on the way to the MIT Media Lab

Make left turn . . . in . . . one . . . point . . . two . . . miles.

—GPS system
on the way to the MIT Media Lab

9

TECHNO-RANT REDUX

Boston. Logan Airport. Right up there with Reagan National, Hong Kong International, and the Bermuda Triangle as excellent death traps. Located on a little spit of land beyond the mouth of the Charles River not far from where the Boston Tea Party took place, Logan is gateway to a city famous for its revolutions; including the digital variety. Which explains why Chip and I are here, deplaning in the elongated upper neck of the country in search of the Massachusetts Institute of Technology's Media Laboratory, a.k.a. The Media Lab, hotbed of digital alchemy. We know it's somewhere in the vicinity, we're just not exactly certain where.

Obviously this a job for GPS!

Okay, so we're gluttons for punishment, but again, this is an adventure, right? We *have* to get in harm's way. No guts no glory. Besides, if ever there was a city that could put a geopositioning satellite system through its paces, it's Boston, land of looping streets and deadly intersections. So we decide to put ourselves, and the GPS system, to the ultimate test: Direct us around the Boston metropolitan area. Guide us through the maze of old cow paths and Indian trails that have been transformed into highways, roundabouts, and cloverleaves to the acclaimed digerati with whom we have come to mind-meld. This time we will not be done in by its opaque and elusive interface.

We head off to the rental agency.

I walk to the counter. "We'd like to rent a car . . . outfitted with a GPS system, please."

Clerk: "Yes, ah . . . (glancing at license), Mr. Shatner."

Standing there, awaiting my next digital fiasco, I suddenly flash back to another time when I was victimized by the unfriendly interface of a futuristic technology.

Let There Be Light . . . or Not

A few years back—no doubt because James Kirk is well known for "interfacing" with talking computers—I was approached by a company that offered to install a computerized, voice-activated, whiz-bang lighting system in my house. The idea was that this would make my home smart, or at least smart*er*. As we all know, no one wants a dumb house. So I took the company up on its offer.

Never again, I was told, would I have to flick a light switch. When I walked into a room I would say to the lights, "On, please." And the lights would come on. When I wanted them out, I would say so, and out they would go. The system would do other things too. Not only would I converse with my lights, but my lights would converse with me by acknowledging my commands.

At least that was the plan.

After many hours of work, after embedding a hive of sensors and chips and microphones and speakers in the walls of my home, the system was installed. I didn't delude myself into thinking I was back on the bridge of the *Enterprise* or anything. I didn't try to have my lighting system call up all pertinent information on the planet Sigma Iotia II, for example, but I have to admit I thought this was pretty cool. The lights, my smart house, and I, would get to know one another. We would communicate as easily as I did with other humans. I would think these thoughts, then make these noises with my throat and pharynx and lungs, and the house would miraculously understand that those noises meant, "Let there be light." Tiny, invisible switches would be flipped without the benefit of any manual labor, electrons would flow, and, like God, my command would banish darkness.

Well, no.

Instead, I walked into my kitchen or living room and commanded the lights to go on, and . . . sometimes they did and sometimes they didn't. Sometimes they talked back to me, but was it in a nice female voice like the ship's computer? No. Instead it carped at me in a hideous, high-pitched, thoroughly mechanical voice that would make Robby the Robot cringe. And when this marvel of modern electronics did deign to speak, what words issued forth from its tinny little speaker?

"Yes, master."

Master! Master?! The first time I heard this I was standing in my kitchen and I wondered out loud, Didn't "master" go out of style in the nineteenth century? Or does this system think it's a genie and I am Aladdin? If it does, it sure missed a critical part of the picture because rarely is *my* wish *its* command. In fact, most of the time my wishes come out as some strangled plea uttered in total darkness, eyes wide, searching for light. Because rarely is my pitiful lights-on request answered. Normally there is silence . . . and darkness. Except for one evening that finally pushed me over the edge. As usual, I was bowling through the darkness calling out for light when, this time, I smashed my little toe into the living room couch.

"%#@!?!!"

And then, through the darkness, I hear a tinny screech.

"Yes, master?"

If *Star Trek*'s computer had operated like this I could have wiped out whole solar systems with a single misinterpreted phrase!

So these days I live, imprisoned in the Stygian corridors of my own home by the technology I have had so mindlessly installed. Like so many digital technologies, I can't communicate with it and I don't understand how it works. It is, pure and simple, badly designed. I can't even switch lights on the old-fashioned way because . . . *there are no switches! They have been removed!* My house has had a switchectomy and left me in the dark . . . literally.

Is this the face of the twenty-first century?

I sure hope not.

Despite my frustration, I perceive a valuable lesson here, just as I perceived it during my first tussle with a GPS. I have learned (and so have my toes) that creating gadgets that are easy to use is a lot harder than faking them, which is, of course, what *Star Trek* excelled in. The ship's computer always responded so perfectly to me and Leonard and everyone else who communed with it because . . . it was a fake! There *was* no computer, there was no voice synthesizer, no artificial intelligence that processed and understood what I was saying and then pulled up the appropriate response from its immense databanks. It was just an old-fashioned deception. I said my line and then Majel Barrett, who was the voice of the computer (and Nurse Chapel) said *her* line. Throw in some sound effects that connoted the machine's "thinking," a few flashing lights, and there you have it: instant artificial intelligence, a machine you can relate to, easily, pleasantly, without a manual.

Were it truly that easy, legions of scientists and designers would be gleefully dancing around their laboratories. Technophobia would not be an issue. Not one of us would ever have to hit a switch or face an interface again. But we're not there yet, not by a long shot, which is why we are in Massachusetts to explore the digital experimentation performed over at the MIT Media Lab . . .

"Your car is in space E-32, Mr. Shatner. It has a GPS system installed."

"Thank you."

We depart to interface yet again.

Navigating the Big Dig

We get in the car. And this time, unlike during our Silicon Valley escapade, we are as electronically outfitted as an *Enterprise* landing party. I have my brand-new palmtop, Chip has a PDA and his ever-handy laptop; our recorder is ready. Both of us have our cell phones clipped to our belts like a couple of phasers.

Boston, we are ready!

And then we emerge from the Ted Williams Tunnel. The landscape looks like some postapocalyptic world where everything has been leveled. Smashed roads, tilting cranes, craters the size of houses. Have you been in Boston lately? It's a nightmare. Not the whole city, the city is fine; but one of the largest road construction projects in the history of mankind is going on here and Chip and I have just driven right into it. Seven billion dollars worth of shattered buildings, rumbling bulldozers, and obliterated roads. As if Boston roads aren't challenging enough.

Cautiously we press on, as though entering the asteroid belt.

Because we were so thoroughly trained by the GPS during our previous close encounter, we have learned to master the thing's cyclopean interface well enough to punch the Media

Lab's address into it before we left the airport. The soothing GPS voice has directed us up to our emergence from the Ted Williams Tunnel without a hitch. But now, as we entered the maze of new and temporary roads, the machine was undone.

I won't go into all the ugly details but suffice it to say in the next thirty minutes we missed ten turns, broke the law three times, and nearly hit a pedestrian. Had we had phasers we would have happily used them to vaporize three other cars, two eighteen-wheelers *and* their *stupid drivers!* At least once I found myself asking Scotty to beam us up, and, somewhere in the middle of it all was a primal scream. I think that happened when the GPS directed us into a parking garage when we thought it was leading us onto a freeway ramp.

Eventually we burned out the fuse for the car horn.

After a few millennia of this madness we finally got straightened out, but not before being often and severely chided by the GPS. See, when you are using one of these things to navigate, it triangulates between three to four satellites orbiting high above the planet to plot your exact position, and then, based on where you have told it you want to go, it's soothing, halting female voice tells you what to do.

"Turn left in one . . . point . . . two . . . miles."

This is impressive when it works, but of course it couldn't work in this mess of rubble and remapped highways. Everything had changed from when the gizmo had originally been programmed to follow its digital maps. They no longer applied to reality. Sometimes we'd be on a road and look down at the cyclops and the arrow representing our car would be sitting *in the middle of nowhere*. The GPS would tell us to make a turn, but because the space between the turns was so short we would already be past the street before we could do anything. When you miss a turn and get off the route, the gadget doesn't get angry, it doesn't set off sirens or bells or anything, it just calmly says, "*Please proceed to the highlighted route.*" And then when you don't—and of course you can't unless you're driving a helicopter or want to kick the car in reverse and plow into every vehicle behind you—it says, "*Calculating new route.*" Which is just the cyclops's way of telling you how thoroughly stupid you are.

Do you know how many times we heard it say, "*Please proceed to the highlighted route*"? And it was only made worse by the fact that it said it so sweetly and calmly . . . time after time after time!

"We are trying, baby! We are *trying!*"

"*Recalculating new route (stupid human!) . . .*"

But you know, in the final analysis, I really can't fault the GPS system. This was an unfair test, even for an infernal contraption. In fact after a while I grew kind of attached to her, maybe too attached. You see, unlike we did in Silicon Valley, in Boston we doggedly stuck with whatever it told us to do, no matter how many laws we broke or pedestrians we endangered. We didn't seek out any humans or gas stations for help. We were going to sink or swim by the GPS, and, finally, even there in the midst of road construction hell, we did finally manage to make our way out like a couple of blind cheese-chasing mice in a maze.

In the end our little geopositioning oracle led us to the right road, headed for the Media Lab. And once on our way, that's when I bonded with the machine. "*Left turn in . . . one . . . point . . . two miles,*" it crooned. "*Bear right in one mile and stay to your left.*" Here I am yacking away with Chip, and it's deliberating with satellites orbiting in the heavens, plotting our course, taking care of business . . .

"*Freeway exit on the right in . . . point . . . nine . . . miles.*"

And I realize I have already become dependent on this thing. I'm obeying it as docilely as a salivating Pavlovian dog; just sitting behind the wheel like some zombie and turning this way and that, whatever old GPS commands me to do. When I make the correct turn it gives me a nice little upbeat chime, and a little downbeat one when I goof up. I don't know if what it's telling me to do is right or not. But I am following orders. We could be going to Timbuktu for all I know. *"Drive off the bridge in . . . two . . . point . . . three . . . miles."*

Yes, master.

But it's a box, for God's sake, filled with silicon, a few wires, and a dime speaker and I am a creature, a human being, supposedly the acme of evolution, my head filled with 100 trillion synapses all firing away . . . and just like that, the Crooning Cyclops is in charge.

And that's kind of the way it happens with technology, isn't it? It's seductive and embraces us until we get used to it and can't do without it. We wonder how we got along without faxes and answering machines, cable television and cell phones. At first it's a luxury, the domain of early adopters, and then after a while we *need* this stuff. In fact the reason we get so upset when computers don't work is because we've become so reliant on them. If we weren't reliant, we wouldn't give a damn. We wouldn't even know they were broken.

I can imagine Glug and Glog sitting around a blazing fire talking in the cave 100,000 years ago. Glug says, "You know I just can't imagine how we got along without fire, can you?"

"Yeah. Ouch!" One technology leads to another, and the next thing you know, you're being ordered around by your car.

You say, so what's the big deal, Bill? So we rely on technology? Isn't that why we created all of it in the first place, to help reduce the grunt work? You know, the wheel, the steam engine, clocks and cars and medicine. On the whole, they make life better and longer and more fruitful, right?

Yes, yes, but technology cuts both ways. Cars pollute, nuclear power can obliterate many human beings, factory emissions contribute to global warming and acid rain. Am I advocating going back? Well, no. It's a moot point! We rely on these advancements too much to do that. Once a technology is out of the bottle there is never any going back.

A little geopositioning system isn't going to destroy us or the planet. I'm not saying that. But as I sat there, merrily weaving through Boston doing a machine's bidding without a second thought, I realized how easily we can come to rely on the technologies we create. And then I realized that if I can come to rely on this little bit of technology, what about more powerful ones? What happens if we come to rely on artificially intelligent machines to manage our lives and then our corporations, then government operations, and then what if they hit a glitch or contract a nasty digital disease? Or what if they, becoming ever more intelligent, decide one day that they'd rather get things done without having to deal with those cumbersome humanoids?

Am I getting carried away? Maybe, but give it some thought (especially as you read through the next few chapters). The more powerful a technology is, the more we had better think . . . hard . . . about putting it out there in the world. Because, again, once it's out there, we can't stuff it back in the bottle. Makes you wonder. Which is, of course, how we got into this whole mess in the first place.

Now, where are we?

"*You have arrived,*" the GPS says.

Right . . . I knew that.

If you must choose between two evils,
pick the one you've never tried before.

—Steven Wright

10

THINGS THAT THINK

"Sure, joke all you want, Captain, but remember your place. We don't care what twenty-third-century character you played. Because in real life we're walking around Kendall Square every night, young and fresh as spring roses, tops in our class with a whole new millennium in front of us, and you—while *we* are out running the universe—*you* will soon be drooling on yourself in the Starfleet Home for Old Admirals."

I didn't exactly *hear* these words, but I'm reasonably sure they are running through the minds of Neil Gershenfeld's students as they chuckled politely at what I had just said to them. Neil is associate professor of Media Arts and Sciences at the Media Laboratory. I have just walked into his lab to continue my quest for the *Star Trek* technologies that are wriggling their

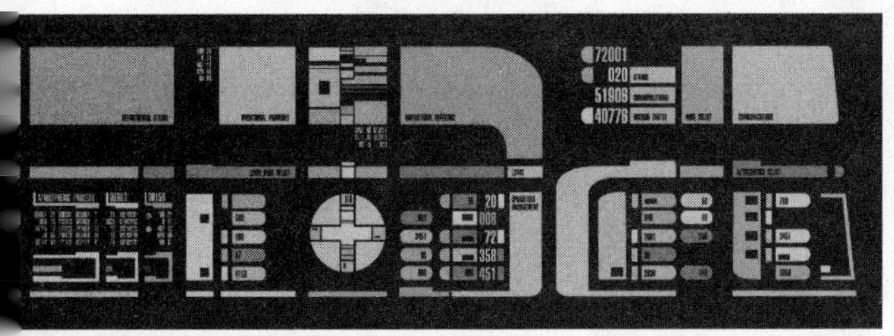

A *Star Trek: The Next Generation* version of an interface with no buttons. *(Doug Drexler)*

way from imaginary worlds into the real one we deal with every day.

Neil himself sits at a long table beneath a tangle of curly black hair, behind an untrimmed beard that grows on his face like a dark, burned bush. The beard tends to focus you on his penetrating eyes. When Chip and I entered only moments before, he grinned, shook hands, and then introduced us to a small group of his students who sat around the table awaiting the arrival of . . . myself.

"My God," I said, scanning their faces, "none of you look any more than twelve years old. Except you." And I singled out one particularly fresh-faced student. "You look eleven." They all laughed, but I knew what they were *really* thinking.

THE LAB

The MIT Media Lab sits in a postmodern I. M. Pei–designed Bauhaus-style building in Cambridge, Massachusetts. Up the street is Harvard University and its fabled ivy-covered towers. Across the Charles River is downtown Boston percolating with commerce and ideas, art and alcohol, not to mention excellent seafood. All around the Media Lab is MIT itself, the Massachusetts Institute of Technology, one of the leading engineering schools on the planet.

It's said that MIT freshmen sometimes long to commit suicide only a few days after they arrive at school because a horrible reality hits them. All of them, you see, were the smartest math and science students in their high schools, but when they show up for their first classes at MIT they quickly realize they are suddenly just one among battalions of freshmen who are as smart or smarter than they are. The shock is just too much to handle. It's called the MIT Syndrome.

The point is the Lab draws on some serious intellectual wattage to do the work it does.

Inside the Media Lab itself more digital concepts, geegaws and gizmos are in development than operate on a Borg cube. Some of the largest corporations in the world shovel large sums of money into the Lab so that the collegial brains within can play, invent, and create whatever their hearts desire. Out of this inventive stew they work to boil up the next generation of future-shaping technologies.

Unlike a university, the Lab is organized across disciplines into consortia that are, in effect, big digital playgrounds, sandboxes that embrace themes rather than specific subjects. There is no consortium devoted to "communication devices," for instance. Instead there is the Things That Think consortium, or the Toys of Tomorrow (TOT) consortium.

The Media Lab was the brainchild of Nicholas Negroponte and Jerome Wiesner. At first sight, you wouldn't expect a man with Negroponte's background to become a visionary purveyor of the digital revolution. He is a classically trained architect who ran the Architecture Machine Group at MIT back in the early 1980s. But he had become fascinated with computers' potential as a tool for designing structures, and felt there was a broader role for them than being monster mainframes or mute workstations that people used as glorified typewriters.

Wiesner, MIT's president, wanted to cross-pollinate work being done in several departments at the university through a laboratory that could integrate computing across multiple disciplines in the arts, media, and sciences. The Lab quickly involved MIT faculty members including heavyweights like Marvin Minsky and Seymour Papert, world leaders in cognition, learning, and artificial intelligence. It also brought in experts from the fields of music, graphic design, video, and holography, and then proceeded to dismantle the walls between every academic discipline in a ferocious attempt to reinvent media and shape the future.

"Sure, joke all you want, Captain . . ."

Irony of life. The older we are, the younger we wish we were. But then when we were young we want to be considered older and more sophisticated. It's an age-old problem. Frankly, given my particular point of view, I tend to agree with Mark Twain. "Youth is wasted on the young." But then I suppose most of the students I was looking at would say wisdom and experience are wasted on the old. Naturally, they have it wrong.

Anyhow, I swear these kids looked like fugitives from a high school science fair. Only a few years ago they probably *were*. But now here they were at a place known for skating the bleeding edge of technology. They walk the hallways of the Media Lab, high-tech mecca, working on projects they had undoubtedly dreamed of as they stared out the windows of their high schools back home only a few years earlier . . . while they silently, effortlessly, aced their physics exams.

Gershenfeld's laboratory is a big room, ceilings twenty feet high, with tables and screens and racks of outrageously expensive equipment strewn everywhere like toys in a kindergarten. Like most other places we had visited it did not exhibit any of the sterile, antiseptic feel of the laboratories that technodopes like me might associate with the future. You know, scrubbed and clean, tastefully lit, not an object out of place. Not here. Chaos was in the air, things were being continually imagined, built, tinkered with, dismantled, and remantled.

I connected with Neil Gershenfeld the moment I arrived because I instantly saw that he was a fellow showman. Don't get me wrong, this isn't to dis Neil's intellectual capabilities. He clearly has a helluva mind, but what I especially liked was the way he *communicated* his thinking. As soon as I arrived I could see he had provided a little bit of theater, complete with stage, audience, and story.

The staging was all digital, of course. (Could it be any other way?) In addition to the standard desktop computer which sat at the end of the table facing him, there was a rectangle of light illuminating the wall behind him—a projection, it turned out, of his own computer screen. An identical projection appeared on the flat surface of the table where he sat. This way everyone in the room could see the computer screen no matter where they sat. Our table was filled with lunch—rows of paper bags stuffed with sandwiches and munchies delivered by a local deli. Before us, like Jesus among his apostles, sat Neil, a smile on his face, ready to get the show on the road.

I hauled a tuna sandwich out of one of the bags and

launched into my standard explanation of what this project was all about.

"We're writing a book which is about the science of *Star Trek* and how so many technologies foreshadowed in the series and movies are now coming true. But, just between me and you, a lot of it is really about my ignorance and my fumbling around, trying to get answers to questions that are pretty basic and fundamental. Who is working on the future, how do they do it, how do the technologies that may shape how we live in the future actually operate? I want to explore the implications of these things. *Star Trek* outlined one possible direction, but now that we are actually creating and exploring these powerful technologies, where is it all likely to lead?"

Like a veteran actor, Gershenfeld immediately took his cue.

"I'll tell you a little about what we're working on. I think it relates," he says. "One of the things we're trying to figure out is how you do computing without computers."

"I like that!"

"Great. Here's an example."

MoMA and the Computer Without an Interface

A year earlier, Neil tells me, popping a potato chip in his mouth, a representative from the Museum of Modern Art (MoMA) in New York came to his lab with a problem. MoMA was preparing a new show on architectural environments. The idea of the exhibit was to explore how our perceptions of public and private spaces are changing. They wanted to include some electronic component, but there was a problem: Terry Riley, MoMA's chief curator for architecture and design, despised computers.

"After all," says Neil, "Terry figures that if people are coming to the exhibit to experience great architectural environments, they don't want to be hunched over a computer kiosk punching buttons and clicking icons. Kind of defeats the purpose, right? You go to the Museum of Modern Art to see art, to be in a beautiful environment and meet other people. A computer is none of that."

So MoMA asked denizens of the Lab if they had any thoughts about ways to use computer technology to enhance the experience rather than get in the way of it. Put another way, they wanted the Lab to figure out how to make the computers invisible. So Gershenfeld and his team huddled with the museum's curators to solve the problem, and after a while they all came up with a very *Star Trek*-ian solution. They made the furniture and physical objects in the show the interface. No buttons, no keyboards, no screens, no kiosks.

How?

Gershenfeld picks a thin metal plate up off his desk and shows it to me the way a magician shows the audience the saw he is going to use to cut his assistant in half. "It's a sensor," he says. He explains that just this kind of plate is also attached to the underside of his desk. He then picks up a couple of coasters. They look a lot like air-hockey pucks. "Think of these as icons, or digital shadows," says Neil. He then puts one of the coasters down on the table and when he does, a display appears on the screen behind him. The coaster has these words on it:

Ghirardo-Cohen House
Clorindo Testa, Architect

Big deal, you say, a table and some coasters. A few beers, some chips and soft drinks and we're ready to watch *Monday Night Football*. Except that in this case the table and the coaster are *talking* to one another. When the coaster is placed on the table, the table understands that it should display the image of the architectural work written on the coaster. And damned if that isn't exactly what it does. There on the screen appears a beautiful rendering of the Ghirardo-Cohen House along with a brief summary about the building itself.

Gershenfeld now moves the puck and places it over a small icon of a blueprint. Like magic, another blueprint, this one full-sized, appears on the wall behind him. Now when he moves his hands to any place on this blueprint, a new, three-dimensional image of the building appears *from the angle you would see it if you were standing inside*. The table not only talks with the coasters, it blithely converses with your hands, understanding where they are in three-dimensional space as you move them over the plans. You have now stepped inside the building, virtually, and can control how you move through the cyberspace with your hands, as if you were a magician.

Pretty impressive. Costly too, you're probably thinking. No, says Neil. That little sensor in the coaster costs one penny. "That's important," he says, "because our goal is to embed intelligence into everyday objects that literally costs just pennies."

All of this means that if you happened to visit the MoMA architectural exhibit, you could interactively explore each architectural work, enter digital versions of the buildings, investigate and learn all about them, yet never once lift a mouse, or click an icon or paw a single computer key. There would be no wires or directions or learning curves. A wave of the hand and you would be interacting with the art, not the interface.

Neil said, "Remember how in *Star Trek* crew members didn't press lots of buttons on the ship's computers, they would wave their hands over the console and things would happen. That's what we wanted to accomplish here."

Powerful Gestures

This is just the kind of faceless interface I sincerely hope will become standard as the new millennium dawns. A world where the simple gestures I would normally use as an old-fashioned human being will not only work in their tried and true, old-fashioned way, but also accomplish magical digital goals not normally associated with them. *Then* I could see how computing could become our friend, not a series of cantankerous struggles.

When it comes to computers, I thought, munching on my tuna, invisibility is a good thing.

Neil's next skit revealed another bit of futuristic-gesture-driven computing that showed how we might bridge the gap between the digital and molecular worlds. He looked at me with an absolutely straight face. "You know one of the most interesting places you can put a computer?"

Well, I'm thinking, I know where I'd like to put *my* computer some days. I smiled. "Where?"

"Your shoes."

"Your shoes?"

"Well, that's not quite accurate. The computer's not in your shoes. The power is."

"The power. Really?" I smile politely, totally in the weeds.

"Let me explain," he says. "We became interested in shoes because they're powerful, literally. When you walk, you generate energy. That energy can power a computer, but not the kind of computer you'd normally think of, not a big box on your desk, or even on your lap. Instead the computer can be very small, even woven into your clothes. And instead of carrying a power supply and a modem cable and all of that, you just wear your special shoes to generate all of the electricity you need."

Fascinating, a generator in your foot.

Now, Neil continues, most of us may not realize it, but we are all electrically charged, head to toe. He places a little electronic shoe on the floor, and asks me to step into it. I do, looking around cautiously to make sure this isn't some sort of nasty experiment normally done with lab rats. Nimoy could be behind it. He's evil and loves practical jokes. Neil wouldn't have cooperated willingly, of course, but Leonard might have threatened him with a Vulcan nerve pinch. In any case, the coast looks clear so I put my foot in the shoe.

I'm not buzzed or electrocuted. No water squirts in my face.

Neil asks me to look at the computer screen. I do. And then he shakes my hand. When he does, information about him instantly appears on the screen. Nothing much, just the sort of information you would see on a business card. You know, Neil Gershenfeld, resident genius and incomparable cyber-guru, The Media Lab. That sort of thing.

I look at my foot and then at Neil and then at the computer screen. "Was there a connection between you shaking my hand and that information appearing?" I ask.

"Well, I rigged it a little, but yes, shaking your hand completed a circuit which made that information appear on the computer screen. You can see how it works."

No, I couldn't. I could see it was cool and interesting, but for the life of me I didn't understand—well, how was it I had so eloquently put it earlier in the meeting—*the long-term ramifications of it.*

In the not very distant future, Neil explains, imagine that you are wearing a very small, low-powered computer that identifies you with whatever information you choose. We already carry this kind of information around with us in our wallets all the time—your phone number, address, occupation, astrological sign, measurements, whatever. Now imagine you shake hands with someone. Your hand then "conducts" that information to another person right through the low electrical charges that pulse through your body. If the person you shake hands with happens to be wearing a similar computer, then the information about you is automatically transmitted to them. Now let's say you also happen to be wearing a small computer display on your glasses, then you'll see the information instantly appear right before your very eyes, a little bit like a thought!

This seemed to me to be the ultimate electronic gesture. Something that combined the digital and molecular worlds in a way that even I could understand. I had read somewhere once that the handshake had probably had its origins back in our earliest days. An extended open hand from one human to another that said, "See, I have no weapon. I mean no harm. Check my hand and see." And by touching, two humans literally established a connection that was open and honest.

Gershenfeld's electronic handshake simply takes the whole gesture to another level. In one simple, age-old motion, yet another kind of connection can be made. Not only a physical one, but an informational one as well. And that informational gesture is also an emotional one, an act of trust as we open, just a wee crack, our minds and our pasts, and reveal a little more about ourselves than is obvious.

This struck me as a good thing. It seemed to be a way that computing could extend our humanity, not blunt it. But didn't this also mean that we were blurring the lines between our technologies and ourselves more than ever? Was it possible that the gadgets that we found ourselves increasingly clinging to, the cell phones and PDAs, would soon find themselves clinging to us instead, like Neil's shoe? Would we soon find the line between what is us and what our technologies are becoming so blurry that they would become indistinguishable? Are we all destined to become cybernetic organisms, variations on the Borg? And what are the "long-term ramifications" of all that?

Yikes!

I was beginning to feel like a character in a bad electronic soap opera. But questions demand answers, damnit, so we shook hands with the prodigies at the Media Lab and departed to explore this digital trend that threatens to close the gap between us and the machines we are becoming so attached to.

The danger from computers is not that they will eventually get as smart as men, but we will meanwhile agree to meet them halfway.

—Bernard Avishai

"Wearable what?"

"Computers."

"*Wearable* computers."

"Yeah."

"Really."

"They're computers that you wear."

"It seems that would be . . . bulky."

"Well, the computers are very small."

"Where would you put them?"

"It would be a little like the Borg, except without the nasty implants, and the zombie gaze. They'd be more stylish. More like jewelry."

"We should check it out."

"It might be worthwhile."

"Where do we go?"

"Pittsburgh."

"Really? Well, rev up the transporter."

"Aye, aye, Cap'n."

—Two futurenauts planning to research this chapter

11

FUTILE RESISTANCE

So I'm touring Carnegie Mellon University, checking out the latest in wearable computers, and I'm figuring that with all of the innovation being heaped upon us that we'd be able to come up with a catchier phrase than "wearable computing." But science is science, not marketing, I suppose. At least you have to admit that it's descriptive.

On the surface wearable computers may sound like just another oddball research project, but when you look more closely at it, the idea kind of grows on you. And after I thought about it, I realized that this was yet another technological trend that *Star Trek* had anticipated. The tricorder—maybe I'm stretching it here because, truthfully, I think Bones and Spock sometimes looked a little silly walking around with that purse-like contraption clutched to their sides as they materialized each week on various planets. But the thinking was right. Wearable.

The Next Generation and other subsequent *Star Trek* series and movies have refined the wearability theme. Communicators are now nothing more than the duranium encased Starfleet insignia on your chest. The medical tricorder is now handheld-sized and entirely new breeds of gadgets have appeared on the scene. Chief Engineer Geordi La Forge's VISOR (Visual Instrument and Sensory Organ Replacement) is probably the most memorable. It's not only a wearable, but a digital organ all its own. (The acronym says it all). Very stylish, too. I was kind of sorry when he received his ocular implants and stopped using the VISOR.

Nevertheless, despite *Star Trek*'s unrelenting prescience, the series has only scratched the surface of the gizmo-laden future that lies around the bend. How do I know this? Because I have been thoroughly, selflessly, exhaustively reconnoitering the future for you and, in the process, have come across people like . . . Dan Siewiorek and Francine Gemperle and their fellow collaborators, which takes us back to Carnegie Mellon University.

Gizmology

Dan is the director of CMU's Human-Computer Interaction Institute and he has his fingers in all sorts of futuristic projects. Francine is the current director of CMU's Wearable Computing Group, part of the university's Institute for Complex Engineered Systems (ICES). Under the auspices of ICES both work with an interdisciplinary team of students and professionals to do research and design on futuristic concepts for high-powered clients like Intel, the Department of Defense, and IBM. So I figure if I'm interested in exploring the gadgetry of the future, I'm in the right neighborhood.

Have you noticed how obvious it's become in the last few years that the world—at least our part of it—is awash in increasingly small contrivances designed to keep us "connected" yet somehow entirely mobile and untethered? At least that's the idea. Pick up *any* magazine and you're eye-to-eye with beepers and pagers; palmtops and cell phones; e-books and MP3 players. You'd think these things were swimsuit models for all of the magazine covers they grace. Wirelessness, it seems, is important these days, and the gizmology that has been developed to make it possible is broad and deep. But is it workable? Or more to the point, is it wearable? Since so much technology seems to be migrating away from the desktop and onto us, wearability, workability, and wirelessness may all soon have to become one and the same.

That's why the experts at ICES have devoted so much serious creative bandwidth to the wearability of computers. In the process they have concluded that as miniaturization hurtles along, we need to do more than simply make more gadgets with which to encumber ourselves. After all, you already have your PDA and laptop and cell phone, never mind your wallet or purse, and maybe a book or a newspaper, the old-fashioned stuff. Enough! We don't have sufficient hands, or Velcro, for more. As it is I hate that lump in my hip pocket called a wallet. Now, thanks to the new gadgets I've acquired, I've been driven to wear a fanny pack half the time. Some of us don't even have fanny packs so we end up fumbling with the things as if they were balls in a juggling act; running from plane to cab to office shifting from PDA to cell phone to wallet.

Got to be a better way.

One solution might be to toss all of these gadgets into the nearest body of water and disconnect (secretly hoping the rest of the world will follow suit). But that's not likely. I'm not detecting any groundswell of anti-gizmo sentiment out there. These days you couldn't organize a revolt without contacting people on their cell phones.

Another approach might be to combine the functions in these gadgets. Phone *and* PDA *and* electronic book *and* MP3 player *and* CD, all in one. But then the next thing you know, you're lugging around another desktop computer and if you take one gadget, you have to take them all because all of them *are* one gadget. There's also the distinct danger that you could end up looking like McCoy with his trusty tricorder slung inelegantly around his neck, and that's not good.

But rather than more, say the folks at ICES, how about if we explore better and smarter. The next generation of thingamabobs, according to Dan and Francine, should feel almost as if they are a part of us; seamless extensions, not hood ornaments. This has resulted in some extremely cool concepts. And those concepts have fundamentally changed my ideas about the shape of the near future.

AT YOUR SERVICE

I've met a few astronauts in my time and they've all told me they've got the greatest job in the world (or out of this world, I guess). But I also know from listening to them that it's tough duty up there. Even now with the new International Space Station finally online, life beyond Earth is a long way from the luxurious amenities the *Enterprise* offers its crew. Astronauts live in cramped quarters, breathing air that would gag a high school jock, in an environment that is nothing like the one for which nature evolved us. No expansive savannas here, no sunsets or sunrises, no place to run.

Schedules are brutal. Astronauts are up at the crack of God knows what—dawn is a moving target when you're orbiting the planet. Circadian rhythms are shot to hell, and sleep comes, at best, in long naps in a bag that hangs against the wall, arms and legs afloat. Add to this the ongoing stress (conscious or not) that the station ship could blow apart like an overinflated balloon at any time, and, well, it's not all glamour and fame. So anything that makes the job easier would be nice.

And that's what a NASA engineer by the name of Yuri Gawdiak at the Ames Research Center got to thinking about. How to apply some *Star Trek*-ian gadgetry to the problem of astronautical stress reduction. The result? Something like a space pet, although that's not how it started out.

Gawdiak's original idea was to provide astronauts with a kind of supercharged PDA. Because of work he had done with astronauts and cosmonauts on the MIR space station he already knew astronauts loved them because they allowed them to record and monitor information wherever they worked. But they said, and this always amazes me, that what they would *really* love to have is something like the fictional tricorder devices we used in *Star Trek*. Small machines that could sense and store large amounts of information, yet be held right in their hands.

Gawdiak worked on that but soon began to extend the idea even beyond anything we imagined on *Star Trek*. First he started to outfit the gadgets for zero gravity. He noticed that crews on missions would often just leave equipment floating around when they weren't using it. Problem was the computers tended to tumble away so they had to keep interrupting their work to reposition them. "I thought they would like a device that would always face them when floating, perhaps something stabilized by gyroscopes," he says.

While he was working on that Gawdiak saw astronauts on a shuttle mission demonstrate toys in weightlessness. He was amazed at how little wind-up, propeller-driven gadgets hopped and blasted around the shuttle with such mobility. Why not, he thought, add a feature like this to the devices he was designing? That would make them not only stable, but migratory. Now he was shifting from a handheld tricorder to something closer to a robot, but this robot didn't need feet or legs because it operated in zero gravity. He called them PSAs (personal satellite assistants).

The next innovation was to personalize the things; load his PSAs with an astronaut's schedule and reminders as well as make them capable of downloading information about experiments. Next came speech-recognition sofware so that astronauts could simply ask a PSA to do something rather than drop what they were doing to punch buttons. Moreover, it would talk back! It might wake you up in the morning, or remind you about your next experiment or let you know it is your nine-year-old daughter's birthday. And it could watch and record with a small camera what you do, which would provide a record of all your experiments for your earthbound collaborators.

When the PSAs aren't busy helping out the astronaut they're assigned to they would patrol the space station, looking for trouble before it got out of hand. It might sniff out excessive carbon dioxide or oxygen (both lethal if their levels rise too high). It might notice a pressure leak far sooner than any human or ground-control instrument would. Or it could routinely plug into and test power and information systems. It might even go "fetch." On one shuttle mission, an astronaut could not find her tennis shoe. When in space where lost time is counted in millions of dollars, you don't want an astronaut spending time hunting around for a shoe. Ultimately it was found, squished under a locker. Says Gawdiak, a PSA could have found the shoe and the astronaut would never have had to waste a single minute of her precious time.

Call me crazy, but Gawdiak seems to have conceived of a cross between a gadget and a dog, custom designed for space travel. Loyal and true, there when you need it, specially bred to record, receive, and transmit information, able to anticipate what you want and bail you out of trouble in times of crisis. When you need to wake up, it digitally licks your face. If there is trouble you can't sense, it barks. If you want to be left alone, you tell it to go sit and it disappears. Rather than feed it you plug it in and recharge its batteries, and, when nature calls, well, nature never calls. This is a creature if ever there was one, smart within limits, far from an android, but made entirely out of silicon chips, capacitors, plastic, and metal. You might call it a domesticated machine.

Think about it. I could imagine something like this being the great-great-great-granddaddy of Data. I mean you have to start somewhere, right?

The Geography of the Human Body

The whole idea of wearable computers is to take portability to the next level. Because you are *wearing* these things you don't have to then *carry* them. Then when you need to use whatever, it's right there, available, or better yet, it's instantly usable without even having to retrieve it.

This is, of course, easier said than done. Wearability implies that what you are wearing is comfortable and doesn't stick out like the forehead of a Talosian. The idea is to wear the gadget like you wear a vest or a wristwatch, earrings or a belt. They fit, they don't hang on you or protrude or make women say to you, "Hey, is that an anvil in your pocket or are you just really connected?"

When I toured ICES Francine Gemperle gave me the low-down. "We've focused our research on locating, understanding, and defining the spaces on the human body where solid and flexible forms can easily rest. We want wearables to be comfortable."

In other words they are putting a lot of effort into making sure that tomorrow's personal, wireless computers aren't a pain in the . . . well, you name your preferred sector of the anatomy. To do this ICES has studied countless wearable objects past and present, from aboriginal jewelry and headdresses to pocket watches. They've checked out sports equipment, gun holsters, medical devices, eyeglasses, and monocles.

There's no difficulty finding things that we wear. Clothes and jewelry are ancient and ubiquitous. But wearable *computers*—they not only have to be inconspicuous, they have to *operate*. This means meeting the ultimate form-follows-function challenge. Not only do you have to understand the terrain of the human body, but you also have to design a device that can rest on that terrain and still be dead simple to use. And on top of that, the geography of any given human form changes from person to person, sometimes drastically (consider the difference in the forms of say, Pamela Anderson and Rush Limbaugh). And finally keep in mind that *all* bodies move. Very rarely do we sit still, and movement constantly changes the shape of anybody's body, no matter what its form.

In a nutshell this means that ICES is trying to hurdle all sorts of nasty obstacles. They must (1) devise very small machines that (2) not only do their jobs, but (3) don't make you uncomfortable no matter what your shape, (4) no matter which way you move.

"It's hard," says Francine, a master of understatement.

So in addition to studying other wearables for ideas about what might work, the ICES crew decided to study the human body itself. Not a bad job if you can get it. Their purpose: to thoroughly understand all of the curves of the human anatomy; learn what happens to them when they move; find out what parts are sensitive to weight, and what areas seemed to have room for computing. They found that certain parts of the body really don't seem to mind having additions attached, and they found that other parts of the body abhor it. For example, the thighs and shins and small of the back turn out to be good candidates for wearable computers, but we really hate having things hanging off of our heads. And, of course, we like to have our hands available to do what nature intended them to do—operate TV remote controls.

So to deal with the way our bodies flex and bend and twist, ICES researchers came up with a few interesting solutions. Some approaches were just old-fashioned adjustable straps, like the kind you find on camping equipment. But the solution I liked a lot was the foam "pods," that gently grip various parts of your body. The thigh pod, for example, clips onto the side of your leg. You can shape the form of the pod however you need to, and only a small strap wraps around your leg to keep it in place. This way you can be certain the pod stays on, but it won't get in your way when you cross your legs or walk. The same with arm pods for the triceps or shin pods located beneath your knees. (Of course you wouldn't wear *all* of these at once unless you were RoboCop or something.)

The idea is that eventually slimmed down, superlight versions of the pods might be where your future wearable computers would live, never to be noticed. Or the pods might somehow become the computers themselves, laced with chips, information, and strange digital abilities, all conveniently attached to you.

Truthfully? The pods are not only comfort*able* but also kind of comfort*ing* like a gentle hug. They don't hang or pull. They caress. Kind of an odd feeling, really, being caressed by your favorite components, but then it beats some other approaches out there.

Robosapiens

Steve Mann is a former MIT student, now a professor at the University of Toronto's Department of Electrical and Computer Engineering. Mann was featured in an article he wrote at the turn of the millennium about himself for *Technology Review*, MIT's "Magazine of Innovation." He's gotten a lot of attention elsewhere as well as being a wearability maven. He confesses, "People find me peculiar." And no this isn't only because he's Canadian! People think he is peculiar because he considers himself a cyborg (cybernetic organism). To prove it, he runs around all day wired to the teeth, his senses augmented to the tune of several hundred megahertz of gadgetry. To engage himself in his cyberworld, Mann wears a big pair of sunglasses that bear a fleeting, if clunky, resemblance to the supercool wraparounds that Tom Cruise has made famous.

Inside these glasses, Mann (ironic name for a cyborg) connects with the Internet, sees with "eyes" located on the back of his head (a camera), monitors his heart rate, even blocks out billboards and ads he'd rather not see. When he goes shopping he shares the experience digitally with his wife who joins him, virtually, from their home or her office to look over the produce and help make the right selections. Frankly I think this takes the connected nature of matrimony a little too far, but I suppose it's a way to eliminate at least *one* argument for the evening.

"I live in a videographic world, as if my entire life were a television show," says Mann.

Is this a good thing?

Anyhow, to spread the word and share the experience, Mann even teaches a class at the university in what I call cyborgology. (The university calls it "ECE 1766: Personal Imaging and Photoquantigraphic Image Processing." How would you like to explain *that* one to mom and dad?)

I'm not knocking Mann's cybernetic efforts. A bedrock *Star Trek* tenet that I absolutely subscribe to is: It takes all kinds. However, I think Steve has made a pretty serious lifestyle choice that's a long way outside the middle part of what the rest of us are looking for. I don't see mainstream America, or even mainstream Canada or Europe for that matter, adopting the bulky confabulation of processors and sensors and switches that he piles on himself every day. Nor do I see people lining up at Radio Shack to buy devices that will enable them to digitally share their produce choices with their significant other.

Even Mann's own students say the equipment they've been provided with is bulky and uncomfortable and a long way off from wearable. Of course they don't use the same system that Mann uses. Instead a company named Xybernaut has outfitted them with commercially manufactured devices, like the head-mounted display that makes them all look suspiciously similar to the Borg. According to one student, "This [head-mounted] display isn't what I thought of as being completely wearable."

Over the long haul, I can't say whether Steve Mann is on the right track or not. He may be. They laughed at Alexander Graham Bell and Robert Goddard and they put Galileo under house arrest. Far be it from me to pass judgment. But I do know one thing, he hasn't solved the comfort problem, and he certainly hasn't solved the "style" issue. Because there is one more problem with wearables that will have to be addressed before the rest of us embrace them. They're going to have to look good. Style is important.

Stylin'

Say what you want about *Star Trek*, but you can't deny that it had a fine sense of style. Always has. Still does. It's a tradition that was instituted by masters like costume designer and consultant Bill Theiss, designer Wah Min Chang, and art director Matt Jefferies; then later maintained by all of the fine artists and designers who have followed in their footsteps on subsequent series and movies.

Part of *Star Trek*'s popularity derives from its cool style, and therein lies a lesson. Even if you can create wearables that are as comfortable as a bed of tribbles and work better than Scotty's warp engines, they had better look good, or forget it. Because when it comes to appealing to a mainstream audience, style is an important ingredient in the recipe for success. Can you say Edsel or Nehru jacket or all of the 1970s? Which is why I like the final phase of the work I saw at Carnegie Mellon's ICES. They call it Streetware.

The Streetware project took the whole concept of wearables, with all of its usability and human anatomy issues, and

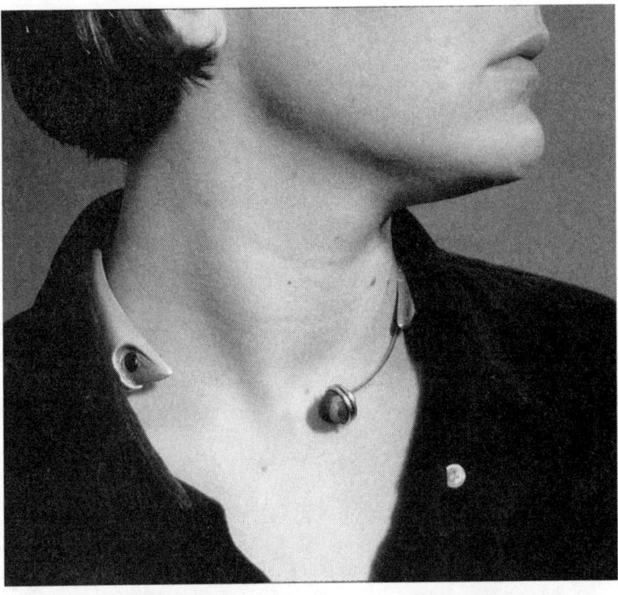

Kneph (not invented by a Klingon). Not only a stylish necklace, but a first-class gadget that can capture your words and fling them wirelessly wherever you like.

laid them out before a gaggle of Carnegie Mellon design students. The goal: help invent the future by developing wearable computers that not only work, but also look good. Since CMU also happens to have produced some of the nation's top industrial designers, applying the talents of design students to the problem of style seemed like a good match.

It was.

Take the personal recording device designed by student Elizabeth Geuder called Kneph. (Kneph? I wonder if Elizabeth is Klingon?) Elizabeth figured that there is a growing number of women professionals out there that could use a way to record ideas, deliver reminders, make mental notes without having to whip out a recorder (even a small digital one) to do it. Kneph solves this problem. It's a necklace with very small microphones attached to each end. Have an idea? All you do is tap the necklace and speak. The thought is recorded and then wirelessly flung back to your own Web site or desktop calendar program where it can then be transcribed into notes that show up on your to-do list. I have a couple of daughters I suspect would find this useful, one that even has a clothing store who could sell it!

My favorite Streetwearable might be Karhu, a gizmo

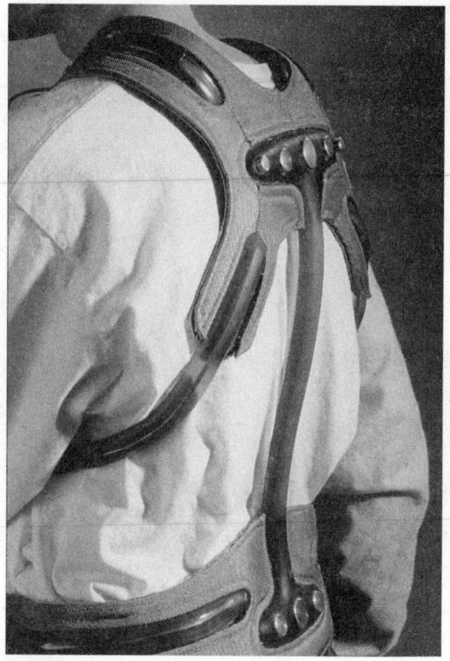

Karhu. This wearable item communicates by nudging you in the right direction. You'll never be lost again . . . at least not physically.

designed by Christopher Kurtz (obviously another Klingon). It's a form-fitting frame for a backpack, but with some very special features. First, Chris envisions the frame being made of "shape recognition material" designed to conform itself to the exact dimensions of your back, shoulders, and waist, automatically. Put it on and it caresses you. (This isn't impossible. Remember the microelectronic materials Andy Berlin at Xerox told us about which can change their shape, depending on the electrical charge they receive?)

A form-fitting backpack frame is a pretty innovative invention in itself; a great way to ensure comfort, but Karhu does plenty more because, remember, it's a computer. The frame also links wirelessly to a network and is outfitted with a geopositioning satellite system. Imagine you've set a course for a day's hiking. The GPS system knows where you are and where you want to go. As you walk the gadget, thanks to its form-fitting material, gently nudges you left or right to make certain you don't get off course!

It doesn't have to have a head-mounted screen or even an automated voice that instructs you on your journey. It just gently guides. I love that! The interface is so simple. I can see something like this being just as useful to vacationers and travelers roaming cities around the world. God knows I could have used it plenty of times. When Chip and I were in Boston doing research we asked for directions to a good seafood restaurant. We wanted to walk; it was such a beautiful fall evening, but getting clear directions was easier said than done. After many labored discussions and several wrong turns, we finally managed to find the restaurant where we promptly stuffed our faces. But what if one of us had been wearing a light version of Karhu under our sweater? It would have simply and silently guided us from the hotel in Cambridge to the dinner table downtown. And never once would we have looked anymore perplexed and stupid than we normally do.

There is one other gadget I have to mention because it addresses specific design problems I've noticed creeping up on cell phone and PDA manufacturers. It's called Stoss, and was designed by Lee-Anne Stossell. Anyhow, Stoss is one of the cooler cell phones this side of the Alpha Quadrant, except that at first glance you wouldn't recognize it as a phone because most of the time it would live on you as a watch. Need the time? You just check your Stoss wrapped there on your left wrist like most watches. But if someone calls you, rather than patting yourself down like a deranged cop or digging through your purse to locate your cell phone, you just unclip Stoss from your wrist because the band is also a small phone. Put one end to your ear, the other to your mouth, and you're in touch!

Stoss. The beauty of this gadget is that, like Kneph, it serves just fine as a piece of jewelry, but it's also cunningly designed to work as a phone *and* Web browser.

Stoss also elegantly solves this other problem I mentioned. It's pretty clear that phones can grow quite small. Just look at the difference between the behemoths we carried around as recently as four years ago and the tiny Star Tacs and Nokias that people whip out around the world today. Now phone makers are going after the PDA market (and vice-versa), trying to figure out the best way to combine today's version of the communicator with something that lets you take notes, send messages, browse the Web and keep track of your appointments. But I have to wonder, even if you could cram a lot of this into one tiny gadget, it still has to be large enough for you to read the information it contains. It doesn't do any good to create a vanishingly small PDA/phone if you go blind trying to surf the Web or read an appointment on a minuscule screen. At a certain point, it just becomes too small to be practical.

But Stoss has a solution. (Other Streetware designs took a similar approach.) The watch part of this dingus also holds a small projector. The information is there, on a very small Stoss screen. In order to see it, you simply project the image onto a piece of paper or a wall. This way you can see all the e-mail or news or Web pages you want.

Smart.

Of course, all of the Streetware designs right now are pipe dreams. Not one of them actually operates and truthfully you couldn't even build working prototypes . . . yet. Stoss, for example, is very small and very innovative, but right now the big obstacle to making it is battery power. (My current, tiny cell phone is mostly battery.) Karhu, on the other hand, will have to await the invention of "shape recognition material," and there are some voice recognition and battery power issues to overcome before Kneph is ready for the working woman. But that's okay because the technology that will make these gadgets possible is on the way. Microprocessors are growing smaller by the minute.

Power? As microprocessors grow smaller they require less power. Soon, as our visit to Neil Gershenfeld's group at the MIT Media Lab revealed, we might be able to power some wearables right from our own bodies!

Which led me to start thinking about the Borg.

The Borg. All networked and not a human thought among them. Fashion is lacking. Not the future of wearable computing, we hope. *(Danny Feld)*

Borgwalk

I can imagine how the Borg might have gotten started this way, a wearable gadget here and there . . . and then somewhere things went horribly wrong. Maybe a descendant of Bill Gates got involved. That would explain the chilling Borg motto, "Resistance is futile." One thing led to another and suddenly you have a race of pale, arrogant cyborgs who have a hard time respecting your personal space.

You can see where I'm going, can't you? We keep miniaturizing computers. First, way back in the 1970s, we get them reduced from the size of a room and placed on desktops. Then they become luggable, then actually portable. Now a lot of them are getting small enough to carry around in the palm of your hand. Soon they'll be so small you'll be wearing them. It's a natural—or at least predictable—progression. At first even these might be a little bulky, but eventually they'll become earring-sized, or embroidered throughout your clothing. As they become smaller they're also likely to become distributed all over your body. A receiver on your lapel, a display screen for your retina, a voice transmitter somewhere else. But even though they will be distributed, they'll be in touch with one another, wirelessly jabbering away like your very own, miniature, all-body World Wide Web. Of course, your personal web will also be connected to the *real* World Wide Web (whatever *that* becomes), and at this point you will have succeeded in becoming part of the bitstorm, with information flowing to you and from you and all up and down you.

I could see where this might lead to borrowing a few of the Borg's tricks before it's all over. I mean scary as it is, is it really that big of a leap to imagine that we will soon not simply *wear* our technology, but *meld* with it? Makes me squeamish, but think it over. A few years down the road, I could see the denizens of Silicon Valley opting to implant a small transceiver right behind their mandible so they can move and shake the world by phone yet still avoid plowing their BMWs into the back of produce trucks on the 101 Freeway. A small tap with their index finger might turn the embedded phone on, a small tap might turn it off. All sound would be directed to your eardrum, and when you reply you would appear to be talking to yourself, which in a sense you are. If nothing else the rest of us could avoid the annoying spectacle of seventeen restaurant patrons simultaneously searching themselves for their phones when a familiar ring fills the air. Just a little ping in your ear, a tap on your jaw, and you've reached out and touched someone.

Nah! Too crazy, you say. We'll never go Borg. Our bodies are too sacred. Well, I hate to be the one to tell you this but implanted computers are already with us and patients with advanced Parkinson's disease have already benefitted. I learned this from my extensive research.

Parkinson's disease is caused when levels of a neurotransmitter called dopamine drop below normal levels. Among other things, dopamine inhibits chemical activity in two specific regions of the brain. When levels are low, chemical activity in these regions get out of hand which in turn causes those who are afflicted to grow immobile. Their facial muscles and bodies become rigid. Ultimately total paralysis sets in and kills the victim. I can hardly imagine a more horrible disease. However, some patients have now been permanently implanted with an electrode that can inhibit the brain's overreactions, and when they are, the paralysis disappears entirely! The electrodes are wired to a small unit in the patients' chests where the whole operation can be controlled with radio signals. When the electrode is activated, the patients are fine. If deactivated, all of the symptoms instantly return.

Similar implants have been used to treat the tremors that diseases like cerebral palsy and multiple sclerosis cause. Cochlear implants—electronic ears—have made it possible for thousands of people to join the world of the hearing. I know of a boy who was almost totally deaf as a toddler. In times past he would have faced life both deaf and mute. Instead he received one of the first cochlear implants and today is a terrific specimen of a teenager; an honor student, and top athlete who converses beautifully with anyone who will listen.

The intermarriage of machine and human is what has made it possible for the great physicist Stephen Hawking to communicate many of the groundbreaking insights he has brought to theoretical physics. In the early nineties the disease from which he suffers even robbed him of his ability to speak. It was already impossible for him to write. If it weren't for the computer system that enables him to form his thoughts from menus of words on a computer screen and then have them spoken through a voice synthesizer, all of the remarkable thinking going on inside of that mind over the years would have been locked away with no chance of release. The loss to science and to civilization would have been incalculable and tragic. So it might be well to keep an open mind when considering the melding of technology and biology.

We're still in the Stone Age when it comes to implants, the experts tell me. For the most part they say doctors continue to treat the brain like a kind of minestrone, adding ingredients and spices here and there in the form of enzymes and chemicals hoping to get the recipe right. But more and more we are treating the brain like the complexly wired invention that it is.

More digital augmentation is on the way I've been told. I came across an article about an artificial retina being developed at Harvard Medical School that connects directly to the optic nerve and is powered by the sun! Silicon "neurons" are being worked on that might help amputees or the victims of spinal injuries operate artificial limbs or recover from paralysis. In a recent experiment researchers implanted ninety-six minuscule electrodes into the brain of an owl monkey. The electrodes collected and analyzed the monkey's brain signals and translated them into digital commands which then winged their way over the Internet six hundred miles away where they moved a robotic arm. In other words, simply by thinking the monkey moved the arm! You can't get much more connected than that! The hope is that someday implants like these will help handicapped people to move damaged or prosthetic limbs just as they move any other part of their body.

Pretty amazing.

The Turbocharged Self

I suppose it makes sense that we're focusing our first uses of digital implants upon undoing the damage that disease and injuries do. After all, we've been using artificial replacements for as long as we've recorded history. There's Ahab's peg leg, a poor replacement for the original, but it worked. And Ben Franklin's bifocals, an enlightened bit of genius that solved two eye problems at once. And today there is everything from wheelchairs and ventilators to contact lenses and hearing aids. I don't hear much debate about whether this is right. Who among us wouldn't use whatever technology we could to restore a loved one or ourselves to a normal life?

But once we have grown more adept at digital repair, how long before we use electronic prostheses to augment the perfectly fine bodies and minds and senses we already have? That's the question on the table now.

Is the age of the turbocharged self upon us?

Dan Siewiorek will tell you that digital technology, based on systems now being explored, may soon deliver eyes that see in the dark or ears that can hear things generally reserved only for dogs. Some futurists believe we'll tend to turbocharge ourselves by slowly internalizing the kind of thingamabobs we already carry around. Why pack binoculars to the ball game, for example, if the right digital implant can simply transform your eyes into telescopes? How about an infrared chip that lets you see heat? They might become standard issue equipment (or are they senses) for every fireman, policeman, or soldier.

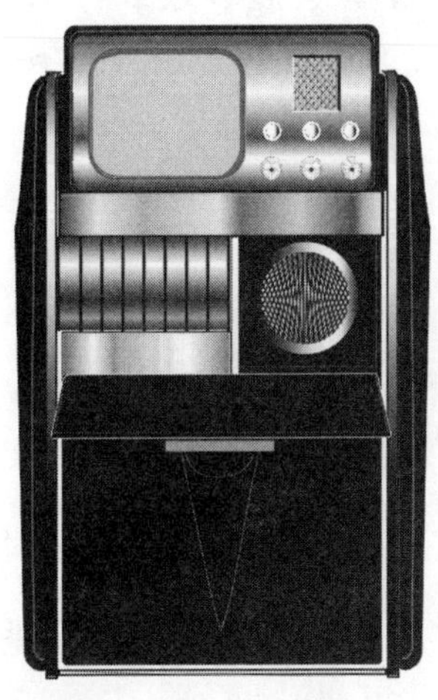

The original tricorder. Wearable, but did Bones really feel chic carrying it around? *(Doug Drexler)*

Chip and I got talking about this and all sorts of possibilities started coming to mind. Our phones and computers and PDAs connect us not only with one another in the real world, we realized, but with information and media in the cyberworld. So how about an optical chip that lets your eyes tune into the World Wide Web, which will soon become a full-blooded, three dimensional world of its own. Why download music onto a computer when you can have it play on demand in your head? And don't forget the mandibular phone!

Or take it another level if you can stand to. Let's say you're aging and your memory needs tweaking so you arrange an implant to keep all your memories bright and shiny. While he's in there, says your smiling Beverly Hills brain surgeon, he can also increase your memory and thinking speed a few extra percent, basically bumping your IQ up several points. Special offer, just for you. Okay, you say, augment me. Next thing you know, you're remembering prom night, the ingredients to your favorite pasta dish, and everything you needed to know to complete the report the boss has been bugging you about without having to take the time to check a single fact. This information was always there in your brain, it's just that now the implants make it easier to retrieve. There's not one detail about a single *Star Trek* episode or movie that you can't remember. Not only that, you're recalling whatever you want to recall instantly and flawlessly. "Uh" and "um" have been removed from your vocabulary. You may not be turbocharged yet, but you're certainly picking up speed.

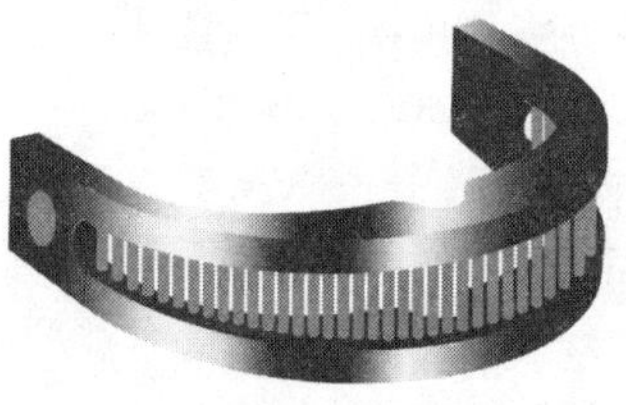

VISOR (Visual Instrument and Sensory Organ Replacement). Pretty stylish for a visual organ. In the future, though, the technology may simply be built directly into us. *(Doug Drexler)*

So, now you're sold on the Beverly Hills Neural Implant System. Next you sign up for the Encyclopedia Britannica memory enhancement chip, and then a new visual cortex which enables you to see at levels of detail and at speeds you never could before. It even comes with instant replay!

If we set off down this self-augmented

road (and I think we already have), will we become the Borg? Is resistance *really* futile?

We may become cyborgs in the twenty-first century, even morph into digital versions of ourselves. I can't say. We may step between the real and virtual worlds with a blink of our electronic eyes, communicate by way of wireless, digital telepathy; enhance our senses and bodies and brains to the teeth; but I don't think that we'll become the Borg. We're just too stubborn. Becoming a Borg would be like being on a perpetual party line, with nary a moment to ourselves. Human beings would never stand for that. There are just some things we won't give up. So no matter how much technological enhancement we can get our hands on, ultimately we'll refuse to erase the boundaries that separate each of us from the rest of the universe, and from one another. Becoming a Borg would mean giving up our personal freedom, and all of human history clearly indicates that we really don't like doing that. The scariest thing about the Borg is that it never has an opinion. Think about it. Can you imagine a human being without an opinion? If there were such a thing, you'd have to give it another name, and whatever it was it wouldn't be human.

All of this scares the bejesus out of me, but what can I say? Am I shaping the future? No. Maybe we're just too balled up with defining ourselves in terms of what we're made of? Metal, carbon, silicon, skin and bones, what difference does it make, as long as the behavior, the essence is still the same. I don't know. I haven't figured it out yet, but hang on, we're about to meet thinkers who *are,* and, being human, they have no shortage of opinions at all on any of this.

PART THREE

ALIENS AMONG US

This program has performed an illegal operation.

—Twenty-first century computer error message when it can't solve a problem

Working . . .

—Twenty-third century computer when trying to solve a problem

In which we learn the nature of intelligence, the power of the brain, and how we may be creating our own replacements. Is the human race on the endangered species list? The history of AI, and the tribble effect that may turn *Homo sapiens* into *Robo sapiens.*

Artificial Intelligence: the capacity of a digital computer or computer-controlled robot device to perform tasks commonly associated with the higher intellectual processes characteristic of humans, such as the ability to reason, discover meaning, generalize, or learn from past experience.

—*Encyclopaedia Britannica*

AI is the attempt to make computers do what people think computers cannot do.

—*Douglas Baker*

. . . computers in the future may weigh no more than 1.5 tons.

—*Popular Mechanics, 1949*

12

AI-OLOGY

Okay, the bitstorm is descending and computers will soon become so woven into our lives that they'll produce a kind of digital ecology as complex as the natural one in which our spear-chucking, loin-clothed ancestors lived eons ago.

That's what we have learned about computing so far.

But this is just one direction the electronic future may take. There are others—one, for example, in which the machines become, not simply small or invisible, but familiar. So familiar, in fact, that they will act like . . . us. In the field of science, this represents the quest for artificial intelligence or AI, one of the holy grails of modern research, right up there

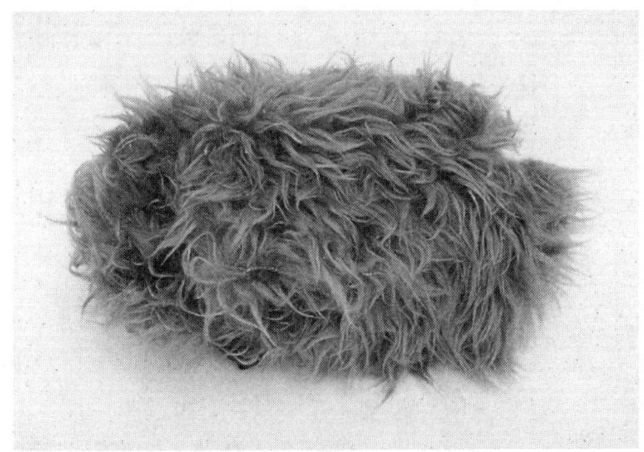

The Tribble Effect: The speed and capacity of machines to process information has been growing at an exponential rate, kind of like tribbles.

with things like the origins of life and the explanation of why Spock's ears are pointy.

Even I, the anti-geek, am familiar with the term artificial intelligence. After all, it's a mainstay of science fiction in general and *Star Trek* in particular. You know, smart, sometimes diabolical, creatures that work and "live" among us. The primary difference between them and us is that they haven't entered the universe in any way we might call natural (screaming and crying through a birth canal, for example). Instead these creatures have been constructed. They are mechanical, unnatural, yet fully conscious, or at least very intelligent. They are also usually more powerful than we would rather they be. There's Gort in *The Day the Earth Stood Still* and Robby the Robot (his stage name) from *Forbidden Planet;* not to mention the Terminators in *The Terminator* and *Terminator 2: Judgment Day.* And who can forget the singularly focused and chillingly calm HAL 9000 from *2001: A Space Odyssey?*

And these don't even include *Star Trek,* where I have personally grappled with so many smart-aleck robots, computers, and androids over the years that I'm sick to death of them. If I see one more threatening capacitor, one more menacing diode, I'm going to rip its head off! Remember V'Ger in *Star Trek: The Motion Picture,* hell-bent on the destruction of pretty much everything? And the M5 multitronic computer created by Dr. Richard Daystrom which was supposed to replace my crew ("The Ultimate Computer"). That SOB had the gall to relegate me to the status of Captain Dunsel (a Starfleet midshipman's term for a part that serves no purpose). But I had the last laugh. When it started acting up and nearly wiped out five hundred Starfleet personnel, I outsmarted it and it self-destructed. (Wish I could do that with my "smart" lighting system.)

Of course not all AI encounters have been entirely unpleasant. There was Rayna Kapec. I fell in love with her in the episode "Requiem for Methuselah." Unfortunately she short-circuited, emotionally, and I never had the opportunity to find out if we were well, compatible. I mean, it's a starship captain's job to be thorough, right?

And then there were the stunning android women in "I, Mudd." They were wonderful to look at but, as Harcourt Fenton Mudd had already learned, they had a funny definition for "serve." It really meant control.

Obviously imaginary artificial intelligence can be terrifically entertaining, and science fiction writers and filmmakers have been having a field day for years with them, but are we anywhere near creating the real thing? Well, we're making progress, but it's pretty clear that conjuring *fictional* intelligent machines is a whole lot easier than building real ones. The AI experts we tracked down often made it clear that as engineering problems go, building an artificially intelligent creature makes constructing a warp drive look like a Tinker Toy project. Why? Because to create an intelligent, sentient machine, you have to somehow manage to artificially accomplish in a blink what nature has taken billions of years to attain in the form of the human brain. No mean feat as I discovered.

The Web in Your Head

If you think about what it takes to think, it's almost more than your own mind can handle. In fact some scientists wonder if it is possible for the brain to comprehend its own complexity. But whether it can or can't, no one can really argue that the wiring that makes it possible is truly miraculous. How, for example, do we organize each of our individual views of the world? Why and how do we feel all that we do? How do we recognize and then solve the problems that we face? Where does the horsepower for all of this come from and how can we best use it to convince members of the opposite sex that we are cool?

Well, it seems to come down to a lot of very small things adding up to more than the sum of their parts. This is what the experts have said. The human brain houses about 100 billion neurons—specialized brain cells optimized for the communication of small bits of information. Each of these cells makes an average of one thousand connections each. That means that there are a grand total of one quadrillion (1,000,000,000,000,000) links, all inside your head. It is the incessant percolation among all of these connections that enables us to see, hear, walk, feel, reason, remember, learn, talk, smell, smile, cry, feel joy, despair, hope, love, fear, and hate. From the human brain have emerged every thought, poem, invention, and insight we know of; the art and innovations of Leonardo, the poetry and plays of Shakespeare, the madness of Hitler, and the philosophy and insights of Christ, Moses, Buddha, Mohammed, and Groucho Marx. From the human mind have spilled maps, movies, mayhem, clocks, the wheel, radio, fire, television, mathematics, rockets, medicine, computers, *and*, oddly enough, the silicon engineering that is so rapidly rearranging our world today and which might make creating creatures as smart as we are possible in the not very distant future.

But why would we want to do that?

One opinion is that artificially intelligent machines represent the ultimate tool. Tools, whether they're a plow, a cell phone, or a robot, are designed to make our lives easier, reduce the workload, and extend our abilities so we can do more, better, faster. Well, what would be a better tool than a robot or android or computer that is our intellectual or physical superior, yet docile enough to do precisely what we ask?

Thinking about these questions made me wonder how we even came to consider the concept of artificial intelligence in the first place. So I checked into it. It's fascinating.

A Brief History of Artificial Intelligence

The Oxford English Dictionary's definition of a computer in the late nineteenth and early twentieth centuries reads like this: "One who computes; a calculator, reckoner; a person employed to make calculations . . ."

Obviously our ideas about computers have changed over the past fifty years. It never occurs to us now that back in the 1940s computers weren't wired up boxes with monitors, but humans—living, breathing people who were literally given a set of written instructions that they followed step-by-step to solve mathematical problems, not unlike the way digital computers are programmed today. The humans didn't have to be great mathematicians themselves, nor for that matter have even the least grasp of mathematics. All they had to do was mindlessly follow the rules that were laid out. If they did, they would get the right answers.

During World War II a lot of these human computers were enlisted to calculate numbers for ballistics tables so the military could determine where a shell would land if it was aimed in a certain direction, at a certain height and had a muzzle velocity of . . . whatever. Toward the end of the war, early electronic computers in the United States were developed for this very purpose. The beauty of these newfangled machines was that, slow and cumbersome as they may seem today, they could calculate numbers faster than Rain Man.

It didn't take long before "computer" stopped being a job description and started being a machine.

The Birth of Computers

In 1940 I was personally spending most of my time in my home town of Montreal just trying to coolly ride a bicycle and keep my cowlick down, but I do recall that things looked pretty bleak for the free world. All of Western Europe from Czechoslovakia to Belgium and France were under Nazi control. The Third Reich was moving east on Russia and fascism was flourishing around the world. Great Britain was at war with Germany which meant that Canada, my home, was as well.

As if this wasn't bad enough, 220,000 British "tommies," the remnants of the British land forces fighting in continental Europe had been cornered in a small French port called Dunkirk, their backs against the sea, several blitzkreiging German divisions bearing down on them. I remember sitting riveted to our old Philco radio listening to newscasts as the drama unfolded. Like the Borg, the Nazi juggernaut seemed irresistible.

Then in a stunningly brave rescue effort, every available ship in England from full-sized troop ships to bobbing dinghies were dispatched to Dunkirk. After three weeks of exhausting and heroic effort, the Brits pulled off the impossible—they saved every tommie in Dunkirk, as well as another 100,000 free French troops.

It was one of the human race's better moments.

The rescue, however, only seemed to postpone the inevitable. The English now had to deal with the German Luftwaffe which Hitler was unleashing to take Britain out of British control for the first time in a thousand years.

At least that was Hitler's plan.

Winston Churchill, however, had a different view. He knew England needed to gain every advantage possible in its struggle to survive so even as he negotiated ferociously to bring the United States into the war, he organized Britain's top mathematicians and engineers and gave them an unvarnished order: Break the German military code. If the code could be cracked, the British military would be able to anticipate and defeat Germany's air invasion.

Just a small job. No pressure.

Luckily, England had just the man to handle the work. His name was Alan Turing, and a more eccentric character you can't imagine. He was a true thoroughbred—skittish, reclusive, and a lethally intelligent mathematician. His brilliance was indisputable, and he quickly became central to developing the mathematics and hardware that enabled the British to crack Germany's infamous "enigma machine" military codes. His efforts during and after the war made him the great-granddaddy of artificial intelligence.

It all starts with him.

To ensure that Turing and his retinue of fellow geniuses could do their work undisturbed, Churchill had them sent to a small country town of Bletchley Park about fifty miles north of London. There, in utter secrecy, among the sheep and cattle, they began work on the "Ultra Project." Supplied with enigma machine plans stolen by the Polish underground, Turing and his fellow scientists rapidly designed and then actually constructed the precursors of the world's first working computers, devices capable of devouring and deciphering German military code far faster and far more reliably than any human possibly could.

These eventually evolved into true computers called Robinson and Colossus which went on to help crack German naval and tactical codes and saved uncounted lives in the Atlantic, Mediterranean, Africa, and Europe. At the height of the war Turing's machines were translating up to two thousand encrypted German signals a day, including messages from "Der Führer" himself.

Thanks largely to Turing's work the Royal Air Force won the Battle of Britain, the Allies won the war, and Earth found itself with a new kind of tool destined to shape the future in ways no one could possibly have anticipated . . . except, perhaps, Alan Turing himself.

Manufacturing Intelligence

Turing saw a future for computers that was a good deal more ambitious than their early, code-breaking days suggested. For security reasons he wasn't allowed to talk much about the plans, so he had to bite his tongue until 1950 when he published the now-famous essay "Computing Machinery and Intelligence." It outlined what was then an outrageous view.

If machines could do abstract mathematics, he wondered, wouldn't it be possible to also eventually create an intelligent machine? Or put another way, could machines be made which could manufacture intelligence "artificially," like a factory manufactures iron ingots, or an adding machine manufactures the sums of columns of numbers? Turing thought so. In fact, he predicted that computing power would accelerate so rapidly that by the year 2000, computers would exist that were as intelligent as we are. (As my home lighting system has proven, he got this part wrong.)

Turing's vision met with some skepticism. How, for example, would we even know if a computer was truly intelligent? (It's obviously easy to spot the stupid ones.) To answer that question Turing developed a test he called the "Imitation Game." Here's how he imagined playing it:

You sit down in front of two computer terminals. One is connected to another human, me, for example, and the second is connected to a smart computer. Now, you start "conversing" with both terminals using the keyboard (remember this was the 1950s; no voice recognition). You're chatting away, chatting just as naturally as can be, regaling me with all sorts of hilarious, charming stories about the office Christmas party, or the time a particular performance of mine moved you to tears.

Now I, being an accomplished raconteur myself, would have no problem carrying on this conversation, but the real computer would have a helluva time because acting human is hard, especially if you don't happen to be one. However, Turing reasoned, the day that it became impossible to know which terminal was linked to a computer and which to a human, *that* would be the day a truly intelligent machine had been born.*

Turing's speculations laid on the table, in a credible, scientific way, the mind-blowing question: Could a machine really think? As insane as this question would have appeared before World War II, in 1950 it actually seemed plausible because early computers were beginning to appear capable of doing a lot of the things that we mere mortals struggle mightily with . . . like very complex mathematics.

Early computers weren't very pretty or efficient; they were monstrously large, stuffed with vacuum tubes and switches and miles of wires. But when it came to swiftly calculating, they were demons. Even very early machines could crunch numbers much faster than any human, performing about 2500 instructions per second and, as long as one of the tubes didn't blow out, they did it tirelessly and without error.

Even for a mathematical genius, adding, dividing, multiplying, calculating is a monumental pain in the brain. Several AI experts have told me just how slow and inaccurate we humans are at this kind of work. (I can also personally attest to this.) We have trouble memorizing any more than seven numbers at a time (which, apparently, explains why phone numbers have seven digits). Running complex numbers through our heads just isn't what we are built to do. But it was precisely what computers do extremely well.

Because they seemed so obviously good at resolving problems that easily confound us, some folks in the field got carried away. Why not have the computers try their digital brains at

*So far no machine has passed the test, but if you want to check out those that have tried, take a look at these sites on the World Wide Web: http://www.loebner.net/Prizef/loebner-prize.html and http://www.lanl.gov/abs.

some other high-end "thinking"—chess, for example. Smart people play chess, right? Let's see how they do at *that*.

As early as 1947 Turing himself had described a computer chess program, and in 1949 Claude Shannon, the inventor of Information Theory—the basis for all modern digital communications—created an actual working one. Every year throughout the 1950s, new chess programs were created and every year computers improved.

Then in 1956, a new wrinkle. Allen Newell, J. C. Shaw, and Herbert Simon, three AI upstarts at Carnegie Mellon University, decided to see if computers could not only crunch numbers, but solve mathematical theorems—problems that required true, logical "thinking." They sat down and wrote what most consider the first true artificial intelligence program. They called it Logic Theorist. (Okay, so these guys were smart, but obviously not geniuses at coming up with snappy product names.) Logic Theorist and another program called General Problem Solver, developed by the same trio the following year, did something that stunned the mathematical world. They worked out proofs for theorems that had been published in one of the great mathematical works of all time, Bertrand Russell and Alfred North Whitehead's *Principia Mathematica*. General Problem Solver even delivered *an entirely original proof for an important theorem that had never been previously solved!*

Mathematicians and computer scientists marveled. This wasn't just a matter of doing something faster and more accurately that any human mind could *already* do. This program had done something that no human mind had *ever* been able to do. And that made people's hair stand on end.

Could machines as smart as we are be far behind?

Well . . . yeah.

Just ask Marvin Minsky.

Miscalculations

In the twilight zone world of artificial intelligence, Marvin Minsky is one of the true pioneers. For fifty years, from the early Turing days, Minsky has been applying the formidable power of his own intellect to the problem of figuring out how to make machines as intelligent as we are. He read Turing's famous "Computing Machinery and Intelligence" essay and it had a serious impact on him. In 1959, he and John McCarthy founded what became MIT's Artificial Intelligence Laboratory, which has been doing groundbreaking work for nearly half a century. Today at seventy-five, Marvin's still pushing the envelope as MIT's Toshiba Professor of Media Arts and Sciences.

Minsky has the reputation of being the quintessential geek. Rumpled shirts mended with masking tape, thick glasses, a head full of books on everything from number theory to science fiction to philosophy. But he doesn't care. Being on the cover of GQ isn't among his top goals. Figuring out how we think and how to make machines that can do the same, *that's* what he's interested in. He's made a lot of progress over the years, and not many people have so directly and fearlessly tackled the really hard questions about machine and human intelligence.

This hasn't been easy. In fact Marvin will tell you that he and other early AI pioneers seriously miscalculated how tough it would be to build machines that acted even remotely human. They found that the things that are hard for us, like the abstract "thinking" needed to play chess or prove theorems, were relatively easy for computers. What they didn't see was that the seemingly simple, common sense things that any child can do, like enjoying a good story, or telling the difference between mom and dad, was brutally difficult for the artificially intelligent. At first this seemed crazy. After all, how much brainpower does it take to tell the difference between, say, a glass of water and a phaser? How smart do you have to be to have common sense, which is by definition, well, common?

Very smart.

In fact billions of years of evolution had been at work developing those smarts, programming them into us. Minsky calculates that the thing we offhandedly call common sense requires that each of us understand about *30 to 50 million separate things about the world* before we can represent them in ways that let us compare them with other things. And none of these are necessarily (to us, at least) terribly complex. This is a glass of water. This is a phaser. You drink one and melt aliens with the other.

Simple, right. Yeah, on the surface. But beneath the surface? It's not so simple. Says Marvin, we don't have a clue about how common sense works. We do not consciously reflect on how we walk or see or brush our teeth or talk. We don't even think about how we think, not usually. And because we don't, the issue of common sense eluded early AI researchers precisely because it was *so* . . . common.

So, it seemed, we had it backward. The easier it is for us, the harder it is for machines, and vice-versa. Apparently the only thing we *can* be sure of is that it takes a boatload of neurons interacting in fiendishly complex ways to deliver what we call intelligence.

Marvin likes to point out an example of the complexity of common sense. We all know that we can use a stick to push objects, he says, but not pull them. And we all know you can use string to pull objects, but not push them, right? Both a string and a stick are straight and long and thin so *how* do we know, instantly and intuitively, the difference between what they can do? Because, he explains, somehow, using all the sensory and neural tools that nature has provided we absorb the little cues and details that make the difference obvious. The first thing we notice is that a string is not stiff!

Figuring these kinds of things out is what *we* have been programmed to do, by the countless hard evolutionary lessons encoded in our DNA—nature's computer program. The capability is so ingrained, that we literally don't have to think about it. But the most sophisticated computer in the world couldn't do any of this if you put a gun to its head (assuming it had a head)!

Ha! The dummies.

But I know what you're thinking. There are some machines out there that aren't so dumb. In fact some might call them brilliant. What about Deep Blue, the world chess champion? What about that?

Fine, let's talk about chess again, and Deep Blue, a computer that also happens to be the world's finest chess player.

Remember how programs were being written in the 1950s that enabled computers to play passable games of chess? Well, programmers just kept plugging away, designing better and better chess playing programs. The first program that could play a full game of chess was written in 1957. The first human fell to a computer in a tournament in 1967, and then the first grandmaster was defeated in 1977.

By 1996 world chess champion Gary Kasparov found himself going head-to-head with Deep Blue. Deep Blue had been designed by a group of Carnegie Mellon graduate students whom IBM had hired to create a digital chess champion. In the first tournament Kasparov won and the Deep Blue team went back to the drawing board with its tail between its legs.

Human triumphs over machine!

But not for long.

Just fifteen months later, the machine returned, newly enhanced for a rematch. This time the machine won the tournament two games to one after three hard-fought draws. In the final game, Kasparov saw he was going to lose and "resigned" before Deep Blue actually placed him in checkmate. Defeat at the hands (I use the term figuratively) of a box full of silicon and wires was too humiliating for Kasparov, a man who had never lost a multi-game match against an individual opponent in his life!

After the match Kasparov admitted that the machine had actually broken him during the fifth game. Why?

"I'm a human being," he admitted to the press. "When I see something that is well beyond my understanding, I'm afraid."

Spock would have loved this, if he were capable of love. Cold, irresistible logic wins out. As I read about this I realized that ironically the very reason, or at least one of the reasons, that Kasparov succumbed was because he *was* human. He felt emotion, and he cracked. The machine arguably won because, unlike Kasparov, it felt no fear. It had no ego or pride. If it made a stupid move, it didn't even know it, let alone regret it or get angry at itself. It could not reflect on a mistake. It was never distracted. It was always, totally focused. All it did, move after move, was sit there and inexorably evaluate 200 million possible positions per second and look twenty to thirty moves ahead.

Does that make Deep Blue smart? Damn right . . . when it comes to chess. It is indisputably the best player on the planet. But is it intelligent? No way. Says Minsky, "Deep Blue might be able to win at chess, but it wouldn't know enough to come in from the rain." And this from a man who believes that machines *will* someday attain human intelligence.

The point is that raw computing power does not an intelligent being make. I wouldn't trade brains with Deep Blue for all of the ships in Starfleet because Deep Blue was an idiot savant, capable of doing one thing, and one thing only, extremely well. It calculated moves and figured probabilities at a stupendous rate. It was a hell-on-wheels statistical number cruncher, but it couldn't look Kasparov in the eye, or, for that matter, move a rook four spaces across the board, something any five-year-old can do.

So what's the point?

Computers have a hard time doing all of the many things that human brains can do because human brains are much more flexible. Despite losing the match, Gary Kasparov can move on and live his life. He *has* a life! All Deep Blue can do is play chess. And when it's not playing chess, it sits unplugged in a room somewhere with the lights out, literally.

This means we aren't about to create Data, *The Next Generation*'s famous android, any time soon. It also means that we aren't in any immediate danger of creating machines that will put us out of a job as the brightest bulbs on the planet. As far as we know our brains are still the most complex objects in the universe. This gladdens my heart. Computer science has accomplished a lot in the past fifty years, but it hasn't yet succeeded in replacing us.

Does this mean that machines will never be built that will be as smart or smarter than we are? Well, not if you've talked to the computer scientists I've talked to. They'll tell you that it won't be any romp in the park, but they believe that the time has now really and truly arrived when highly intelligent, dare we say conscious, machines are within reach.

Why should we believe them? Why, after so many years of getting it wrong, do some scientists think that this time they are right?

Two words: Exponential growth.

The Tribble Effect

Exponential growth is one of those phrases you hear thrown around a lot these days. It comes up at cocktail parties when people discuss the number of extramarital affairs certain politicians have had, or in conversation when athletes are rattling on about their latest contract. Truthfully? Most of us don't comprehend the real power in those two little words any more than we comprehend operating our videotape machines. I, however, world-class anti-geek, will now attempt, without a net, an explanation.

But first I must enlist the aid of an infamous *Star Trek* character—the tribble.

Double This

Let's say you have two tribbles and an unlimited supply of quadrotriticale (tribble food). Now, as we all know, from "The Trouble with Tribbles," tribbles are prolific little hairballs, and if anything in the universe excels at exponential growth, it's these guys.

Let's say we have two to start with—well fed and primed to procreate. Furthermore, let's say they double their numbers every hour. (That's what exponential growth is all about, continually doubling the product of previously doubled numbers at regular intervals.)

So how bad could it be by the next day, you ask yourself? You have a big house. You can handle a decent sized party. Fine. After four hours you have sixteen tribbles. No problem, you can entertain sixteen humans with a couple bowls of nachos and two cases of beer. Tribbles don't even need that. You just serve up another dish of quadrotriticale.

After eight hours you have two hundred and fifty-six hairballs. Okay, a third of the day is past and it's getting a little tribbling, but you're coping. At least they're cute and furry.

Eight additional hours later, sixteen hours in all, you now have 65,536 tribbles. Now you're wondering what the hell is going on. Even though they're no larger than a fuzzy tennis ball it's getting pretty crowded, and if they aren't potty-trained, well, not a pretty sight.

Finally, twenty-four hours pass. And guess what, you're dead, because you've been suffocated by 16.8 million tribbles, give or take a few thousand. They have filled every nook and cranny of your palatial home. Give them another day and they'll completely overrun the planet.

Headline:

Earth: Tribbled to Death!

That's the power of exponential growth. When you start doubling numbers, and when the numbers get big, serious changes start coming . . . fast. If you graphed it, the graph would look like a hockey stick. At first the trend doesn't seem like much, hardly noticeable, but then, suddenly, things start to change *very* quickly.

As a result of exactly this kind of exponential growth, scientists now believe we are at the part of the hockey stick where the trend starts to rapidly sweep upward. It took fifty years to get here, but we are now doubling some very large numbers. Computers are currently more than *100 million times* faster than they were a half century ago. (By the time you actually read this, they will be approaching 200 million times faster because of the lag between writing this sentence and the publication date of the book. In another twelve months—400 million times faster. You get the idea—Chip.) If the automobile had made as much progress in the past fifty years, a car today would cost one one hundredth of a cent and travel at better than warp 1 (with all due respect to Albert Einstein). On the other hand, I'm guessing that you rarely have to reboot your Ford or Toyota while you're driving it, and I haven't yet had to upgrade my steering wheel.

The Tribble Effect wasn't all that obvious fifty years ago. In fact, back in 1943, Thomas Watson, the chairman of IBM, remarked in a now famously short-sighted statement that there was probably a global market for maybe five or so computers, tops. Five! Thomas Watson was right about a lot of things, but on this one he was off by a factor of several million.

Crazy as his statement seems today, it was actually a pretty sensible observation at the time. In those days, after all, computers were the size of tanks. They were so big, in fact, that the first computer bugs weren't actually badly written lines of code, but *real bugs;* critters that would crawl into the spaces where the machine's vacuum tubes were located. The poor arachnids or insects or centipedes would get fried, *and* short out the computer. Then some sorry soul (probably a graduate student) would have to crawl back in there, "debug" the system, and replace the destroyed vacuum tube. Eventually computer systems became a little more sanitary, but anything that derailed a computer's operation henceforth came to be called "a bug," the bane of every one of us who has ever seen that wonderful message: *This program has performed an illegal operation.*

But it wasn't just the size and complexity of the machines that kept the computer population relatively small in the early days, it was the expertise needed to run them too. Exotic codes and programs had to be written in arcane mathematical languages. Only a few carbon-based life-forms knew how to speak in computer tongues like FORTRAN or COBOL or BASIC. Engineers that could make computers work were the high priests of the digital world, like ancient oracles with special powers. They and they alone could commune directly with the great machines.

Could anyone ever have imagined that a little more than fifty years later, millions of us would have our own personal computers? Or that mobile phones would be as ubiquitous as nineteenth century buggy whips? Or that we would carry around "palmtop" devices that are thousands of times more powerful than those early room-sized dinosaurs?

No.

How did this happen?

I am given to understand (being infinitely ignorant about these matters) that the digiscape kicked into a higher gear when Gordon Moore and Robert Noyce developed the first integrated circuits and silicon chips just about the time *Star Trek* (the original series) was going off the air. That was when Moore, who with Noyce, was soon to become one of the founders of Intel Corporation, made his famous prediction which later became known as Moore's Law. It went like this: Every eighteen to twenty-four months, a new generation of silicon chips would be created that would double the capacity of the generation before it.

Back in the 1950s commercial vacuum tubes had already evolved into transistors, much smaller mechanisms for storing, copying, and manipulating digital information. Then in the 1960s Robert Noyce and Jack Kilby (who was working separately) conceived the integrated computer chip. These were tiny wafers fashioned out of specially refined sand, where engineers learned to imprint ever smaller transistors right onto silicon, not unlike the way a photographic image is imprinted on a negative. (This is Intel's, and Moore's, claim to fame and fortune.)

Basically this meant that every twenty-four months the processing power *and* speed of integrated circuits doubled *at no increase in cost.** This is the kind of growth that the experts came to call "exponential." (Capacity is now doubling close to every *twelve* months.)

Unlike Watson's prediction about the computer market, Moore's about the increase in computing power was right on the money (as anyone who owns Intel stock has learned). Which brings us back to the scientists who believe that exponential growth (The Tribble Effect) means that intelligent machines are just around the bend. And that brings us to another question: Is it the machines that are around the bend, or the scientists?

*Actually, if you expand the concept beyond integrated circuits, it turns out that computing power has been growing at an exponential rate, not only for thirty years, but for one hundred years, going all the way back to the first use of a machine called the "analytical engine" that helped tabulate the United States census in 1900.

The reasonable man adapts himself to the world; the unreasonable one persists in trying to adapt the world to himself. Therefore, all progress depends upon the unreasonable man.

—*George Bernard Shaw*

It's alive! It's alive! It's alive!

—*Dr. Frankenstein*
Frankenstein *(The Movie)*

13

MAD SCIENCE

Someone recently said to me that with all of the strange inventions emerging these days, mad scientists could get a bad name. I told them to wait and see what the next few years had in store. Not even a fictional mad scientist could come up with scenarios as wild as the real ones that lie ahead.

The idea of mad scientists—eyes wide with lunatic insight, hair shot out at all angles as if electrified by the wildly firing synapses beneath their scalps—these characters are woven throughout popular culture. Ever since Dr. Frankenstein—actually, ever since Dr. Faust—they have been a mainstay of literature and entertainment and science fiction. They represent the wonderment disease run amuck.

Being a mad scientist requires special talents, a certain mix of basic, unassailable qualifications. Think about it. Every mad scientist is . . .

1. Brilliant.
2. Just a *little* bit too curious for his (mad scientists are almost never women) own good.

3. Insane, and, ultimately . . .
4. Done in by pursuing things he should have had the good sense *not* to pursue in the first place.

Faust sold his soul to the devil and went to hell in the bargain. Frankenstein (the doctor, not the monster) created life, artificially, and then his creation knocked him off. Dr. Jekyll explored the dark side, courtesy of Mr. Hyde and did himself, and a few others, dirty along the way.

Plenty of other fictional curiosity-mongers have suffered the same fate over the years. *Star Trek* certainly didn't shy away from them. We could stock a starship with mad scientists.

So, you're wondering, what's the point?

Well, I'm not saying that any of the scientists we've been talking with are mad. In fact, I think they are all eminently sane. But I can't help but wonder, generally speaking, about mad science whenever I start hearing about developing smart machines and humanlike robots. It just seems to be dabbling in areas where we are out of our league. You know a mind might be a terrible thing to waste, but it's *really* a scary thing to create.

And I don't think we should kid ourselves—building minds is really what is at the bottom of the field of artificial intelligence. True it may have a lot of other goals as well—smart cars and clever TVs and intelligent Cuisinarts—but in the final analysis let's remember that it's about constructing *a device that is very smart!* An *alien* intelligence. Creatures, in short, that are not human, but are at least as bright as we are.

Now as any self-respecting science fiction fan knows the first aliens we encounter are *supposed* to come in the form of a superior race of bizarre-looking creatures from *another* world, not the one we already inhabit. And they certainly aren't supposed to look and act like us. H. G. Wells, Robert Heinlein, Steven Spielberg, and Gene Roddenberry have told us so. But based on everything that I've heard and seen recently, the first aliens that we meet will very likely *not* be from *out there*. They'll be homemade!

Why do I believe this? Because as we started to explore this subject, we came across the work of two scientists who, we immediately realized, we had to talk to. Not because they are mad, but because they are willing to say things and explore ideas that most other scientists have not been willing to say or explore. Or if others have, they haven't done it quite so eloquently and so publicly. The truth is, these guys have raised a little hell, stirred things up by saying unmentionable kinds of things, just like their intellectual grandfather, Alan Turing. And like Turing they have the reputations to back their thinking up. These guys are serious scientists, with serious credentials.

Therefore bound and determined to go wherever necessary to bring you the facts, Chip and I once again hopped into jets piloted by people we do not personally know and tracked these guys down.

Their names are Hans Moravec and Ray Kurzweil. Fine men. Very bright and thoughtful . . . but they think disturbing thoughts. They believe that machines will someday be smarter than we are.

First we went to see Ray.

Intelligent Questions

Ray Kurzweil is not your standard-issue academic. Yes, he was a student of Marvin Minsky's at MIT, and, yes, he has a list of academic accomplishments as long as your arm, but rather than work within the ivy-covered walls of any particular university he instead struck out on his own into the business world. He's actually something of an anachronism. In an earlier time we would have called Ray an . . . inventor.

Since he was a kid, Ray always wondered about intelligence. He wondered about what it was, and what it might take to re-create it. Maybe this is not what most kids think about, but for Ray it just seemed like an interesting problem. And after a while he came to this conclusion: At its heart, intelligence is about pattern recognition.

Pattern recognition?

Pattern recognition.

Ask yourself this question: Does the world make any sense to you? Well, okay . . . philosophically sometimes it doesn't make sense to me either, what with all of the irrational murdering and suffering we see. But generally, at least on the most basic, put-this-foot-in-front-of-the-other level, the world *does* make sense to most of us, or, probably more precisely, our brains make sense of *it*.

When I get up in the morning, I put my feet on the floor (as opposed to trying to put them on the ceiling) because I intuitively understand down from up. I navigate from room to room without bumping into the walls, usually. When I hear certain kinds of sounds emanating from the radio or television, I recognize them as speech, not music or static, and my mind harvests meaning from them. And when I gaze at the newspaper and see all of the strange markings and images on its pages, my brain makes sense of them and translates it all into information which I then process and log away. I do all of this—we all do—without thinking, without reflection about *how* we do it, yet it is extremely complex work.

How is this possible? How do the billions of neurons folded layer upon layer within our skulls accomplish these amazing feats so effortlessly? Says Ray, we do it by recognizing the patterns of what we experience, seeing similarities and dissimilarities that when taken together help us draw order out of the chaos of sensory input that surrounds us. Solve the problem of pattern recognition and you can solve the problem of intelligence.

The original Kurzweil Reading Machine. It solved the thorny character recognition problem that plagued artificial intelligence researchers back in 1978. Kurzweil's first customer was Stevie Wonder.

My Son, the Genius!

I first met Ray at a little party he was giving one crisp, New England afternoon for friends and business associates. He had invited Chip and me to attend before our scheduled interview with him the next day. It was an informal wine and cheese affair to show off twenty-five years worth of Kurzweil inventions. His mother was there, and she introduced him to me, beaming. "This is my son . . . the genius."

Looking at him I could buy that—a compact man with wiry, gray hair and a confident calmness, almost a shyness, that I imagined might have characterized a younger Einstein. (He combs his hair better than Albert did, though.)

The evidence of Kurzweil's considerable talent and drive was all around. The Kurzweil Reading Machine, the first general purpose, commercial optical character recognition (OCR) machine. It hit the market in 1978. The Kurzweil 250 Music Synthesizer, the first machine to successfully re-create the sounds of a grand piano, electronically (1984). The Kurzweil 3000 Reading Machine, a next-generation computer system that can scan text and images and then read them back to blind or learning disabled people in a voice that actually sounds quite human (1996). There were even a couple of computers run by Kurzweil software that produce pretty good artwork and poetry.

Looking at all of this innovation, I had to ask Ray, "Why?"

"Well, it goes back to pattern recognition again," says Ray. "I specialize in developing software and computers that recognize all kinds of patterns whether they're in sound waves that make up human speech or visual patterns that make up printed language. It turns out that ninety-nine percent of human intelligence is pattern recognition. We're excellent at it. That's the great strength of the human brain. Like I said, it's been an interest of mine going back to high school."

Ray's Got a Secret

A lot of high school scientific fascinations end up revealed in science fairs, usually as tabletop volcanoes or small-time Tesla coils, but not Ray's. His were revealed on national television, on a show called *I've Got a Secret*. For those of you too young to remember this show, first, curse you and your youth. Second, let me fill you in.

I've Got a Secret was an enormously popular television game show way back in the last century in the fifties and early sixties. Basically it was twenty questions with celebrities. Someone would come on the show, like Ray, and state their secret for the host. At the same time, the secret was also revealed to everyone in the studio audience as well as the millions of people watching at home. On stage, opposite the show's host, sat four celebrities who would ask the guest a series of yes or no questions, whatever came into their well-known heads. If they kept getting "yes" answers, they could keep asking questions. If they got a "no," on to the next celebrity

So here we are in the early 1960s, a few years before *Star Trek* even aired its first episode, and a teenage Kurzweil walks out onto the stage, sits down, and plays a composition on an old upright piano. When he's done, he walks over and whispers his secret into the ear of Steve Allen, the show's host at that time.

"I built my own computer," he says.

"Well, that's impressive," says Allen, "but what does that have to do with the piece you just played?"

Again he whispers into Allen's ear. "The computer composed the music."

Lots of raised eyebrows in the audience and around the country. A homemade computer wrote the music? And a kid built it? That's a helluva secret.

Former Miss America, Bess Myerson, asks the first question and she gets nowhere, but the second panelist, humorist Henry Morgan, took only a couple of pointed questions to guess the secret. Wild applause and grins all around.

This was Kurzweil's first brush with notoriety.

"See," Ray tells me, "I built this computer and fed in melodies from a particular composer like Chopin or Mozart. And it would recognize patterns common to that composer and then it would write original music using those same patterns. It was all the computer's work, but the music it wrote sounded like the composer it had been listening to. It didn't have the genius of the original, so it sounded like a third-rate student of Mozart. But it was recognizable. You could tell, 'Okay, that's imitating Mozart.' "

I looked at him. I guess you could say my expression was . . . blank?

He looked back at me, and shrugged. "What can I say? I'm fascinated with human intelligence."

So fascinated, in fact, that a few years after graduating from MIT, Kurzweil invented a machine that could read. The proof was right there in the office where the little party had been going on.

The Kurzweil Reading Machine is a pretty impressive piece of technology, even today more than twenty-five years after its creation. Not only can it scan a book, and recognize the words inside of it, but it can then read them all back in a voice that sounds, if not perfectly human, at least as good as Robby the Robot in *Forbidden Planet*. And this was accomplished in 1976, during the days when most computers were still mainframes.

Kurzweil tackled this problem because at the time one of the classic and unsolved pattern recognition problems in computer science was finding a way to teach a computer to identify printed or typed characters no matter what the typestyle or printing quality.

What's the big deal, you say. I can read all kinds of typefaces and don't even think twice about it. Well, exactly, because what are we great at? Pattern recognition! Once we learn the basics of reading, we can pretty much read any sort of type or font because we get the general pattern of each letter. It can be Goudy, like these letters; ***or Utopia, like this font***; or Franklin Gothic. Whatever the final form of the letter, we easily recognize it. We can even read pretty poor handwriting, unless, of course, it's a prescription written by a doctor, in which case you have to be a trained pharmacist. Not only that, if there are a couple of splotchy or missing letters, it's not a big deal. We humans figure it out.

But that wasn't the case with computers, certainly not in 1978. True, a few systems could read typefaces if they were very specific and fed into the machine under perfect conditions, but a general-purpose computerized reader didn't exist. The trick, the challenge facing Kurzweil was to create a program that could read *any* typeface.

To accomplish this Kurzweil, who always had a serious entrepreneurial streak, founded Kurzweil Computer Products, Inc., and pooled the talents of a few fellow MIT students to get it rolling. Together they wrote the "reading" software.

They solved the thorny character recognition problem this way. You have an ignorant computer. It understands nothing about letters, reading, or writing. It's just an information processor. To this machine the letters A, B, C, D, etc. are no more than a bunch of senseless, random lines. Kurzweil and his team realized that for the computer to be able to "see" a pattern in the lines that make an A or B, regardless of the font, they had to help it learn what was unique to *all* A's or B's. The computer had to learn the A-ness of A.

"So," says Ray, "we taught the machine the unique qualities of each letter's shape. For example, a capital 'A' has this sort of triangular portion on the top where you have a white region completely surrounded by black. We call that a loop. An 'A' has a loop. A 'B' has two loops, but a 'C' doesn't have any because it's not closed. An 'A' has a connection between the north center point and the southeast and southwest points. It *always* has a connection between these two lines. And no matter how you make an 'A' if it doesn't have those basic abstract qualities, the software won't recognize it as an A, or anything else for that matter. We wrote programs to detect these abstract qualities for all letters."

And it worked, brilliantly. That was the good news. The bad news was, so what! Kurzweil's optical character recognition machine was a beautiful piece of engineering. It solved a fascinating academic conundrum, but what practical use did it have? It was a solution in search of a problem. Then during a business flight Ray happened to sit next to a blind man. They had a long talk and, eureka! Ray walked off the plane with the problem his solution had been searching for. He would create a machine that could read printed and typed documents out loud to the blind! All he needed was a flatbed scanner so documents could be easily read into the computer, and a program that could take the type it was recognizing and read it back in a voice that made sense. That's all.

Problem was, neither of those technologies existed. Undeterred, Kurzweil and his team simply went away and created them, and then they fused everything—scanning, character recognition, and voice synthesis—into one machine. And there you had it—history's first print-to-speech reading machine for the blind, a.k.a. the Kurzweil Reading Machine.

Never let it be said that random acts of kindness aren't good business because the invention of the Kurzweil Reading Machine soon led to a collaboration with none other than superstar singer/musician Stevie Wonder. Wonder, who is blind, had stopped into Kurzweil's offices after hearing about a story on the reading machine and became the company's very first customer. Shortly afterward government agencies followed suit.

Later during a conversation with Kurzweil, Wonder asked if there wasn't some way to combine the power of digital music technologies with the beauty of acoustic instruments, like the piano and guitar. Just like that Kurzweil jumped on the problem, and before you could say "Living for the City," the Kurzweil 250 hit the market—the first computer-based instrument that realistically re-created the sounds of a grand piano and most other orchestral instruments. (Where would bands and lounge singers be without them today?)

You Can Call Me Ray

Kurzweil next tackled the problem of speech recognition, another one of those things that seems to come so naturally to us, but is a major pain in the virtual neck for computers. When computers "hear" a sound, they have no context for what that sound is because they have no real intelligence. It's just a noise with no more meaning than a book falling off a desk or the wind blowing through the trees. It's all the same to a machine. So the trick with getting a computer to comprehend—no, wrong word—to *translate* speech, says Ray, is to train it to recognize certain patterns and assemble them, accurately, into the words that those particular sounds represent. This makes recognizing printed words look like a blissful spring walk through the park.

Nevertheless, Ray and his team have tackled the problem and created software designed to understand spoken language. They called it Voice Xpress (sold to Lernout and Hauspie Speech Products in 1997). The software, which is designed to work on your standard desktop computer, has been well reviewed and is selling well.

Anyhow, since I am the world's worst typist and abhor keyboards, I had to try the software out. If I could just talk to that confounded computer in my office, like I used to talk to the *Enterprise*'s onboard computer; and if I could just have it *listen and understand me*, well think of the possibilities!

So I sat down in the conference room at Kurzweil Industries with Ken Linde, the company's systems manager. He set me up with a standard laptop. I placed the nifty head-mounted microphone on my noggin and adjusted it just so, the better to record the lightning insights that would usher forth from my mouth.

"You'll have to talk to the machine for a little while," Ken told me, "until the software becomes familiar with your voice and pronunciation." Well, that made sense. Naturally the software had to learn the patterns of my speech. But once it did, I would never again have to tap buttons on my computer to enter my thoughts.

But, as we know, I have a special talent for confounding machines and software, and it was no different in this case, although I have to say the words that stumped it were pretty surprising.

Now being a great thespian with an unusually mellifluous voice, we all know that I pronounce everything perfectly. So, of course, the problem couldn't have been with me. Nevertheless, every time I said "Ray Kurzweil," I thoroughly confounded the software. "We had a wonderful meeting," I said into the head-mounted microphone, "with Ray Kurzweil, and his . . ." And then on the laptop screen before me appeared: "We had a wonderful meeting with Gray Course Well, and his . . ."

"Ray Kurzweil," I repeated.

"Wave Kurzweil," typed the machine.

"Ray Kurzweil!" I said, perhaps a touch too loudly.

"Breakers-well."

Try as it might, processing like a demon, the software was flummoxed at every turn. In the end it finally settled on recording my spoken "Ray Kurzweil" as "the Great Kurzweil."

Maybe, I thought as we finished up, Ray has that software running pretty well after all.

We left it at that.

Prognostication

Seeing these technologies that we took for granted in *Star Trek* was fascinating, and I could see how, over time, and given a lot of help from some very creative human beings, that computers might likely grow extremely smart in the coming years. But would we soon be dealing on a daily basis with machines that could pass the Turing Test?

Predicting the future is actually something that Ray Kurzweil seems to have a talent for. It seems his work has naturally drawn him into the habit of prognostication, with interesting results. More than ten years ago, for example, he predicted . . .

1. A computer would defeat the world chess champion in 1998. Result? Deep Blue whupped Gary Kasparov in 1997.

2. A worldwide information network linking millions would emerge in the 1990s. Result? The World Wide Web emerged in 1994. You know the rest.
3. Around 2000, computer chips with more than a billion components will emerge. Result? They did.

I don't think even Ray would say that this means everything he predicts will come true on the schedule he outlines. But, given his fascination with human intelligence, and his personal efforts to imbue machines with that intelligence, and given his success at predicting future events as various technologies race to their next level; well, he sure seemed to be a good person to ask the Six Million Dollar Cosmic Question: Will we ever create a machine that isn't simply smart or clever, fast or networked, but also fully conscious and intelligent? Will we someday gaze into the alien eyes of a digital creature so sophisticated that if it commits a crime, the police will have to read him his Miranda rights?

Ray didn't blink.

"Two thousand twenty-nine," he said.

More precisely Kurzweil says that within twenty-seven years, your average thousand-dollar home computer will have the same processing power as your brain.

But if the human mind is so complex and we still don't understand how it works and we don't have a clue about what consciousness actually is, and we sure don't know what other breakthroughs and events might intervene, how can Kurzweil, or anyone, predict that we'll have the digital equivalent of ourselves-in-a-box sitting around within three decades?

It all goes back to the Tribble Effect.

As I am writing this, Advanced Micro Devices, Inc. and Intel have announced the creation of the first 1000 megahertz silicon chip for a home computer (this will be very old news by the time you read this, but it's a big deal right now). These chips will be capable of performing one billion calculations per second. That's five million times faster than the biological circuits in your head. This is fast, right? Right, but even so, as of this moment (and again this is changing all the time) your standard home computer still has only *one one-millionth* of the computing power blasting away in your head.

Why? Because our brains have so many more circuits than any computer currently at work. Even *my* addled brain operates with about 100 billion neurons. (You may remember this is about the same number of galaxies in the known universe!) Even more amazing, each neuron averages about one thousand connections, and every one of those is capable of working at the same time. In computer science jargon, this makes my brain, your brain, everyone's brain "massively parallel." And that's the trick.

Each of the connections within our brains is slower than any one connection in a computer, according to Ray. A neuron performs roughly two hundred calculations per second, *but* there are so *many* connections that no computer can come even remotely close to what our old-fashioned, carbon-based noodles are capable of.

However, at least when it comes to raw computing power, this is going to change, *big time*. If you do the tribbling math, says Kurzweil, by the year 2020 an ordinary home computer will have eliminated that million-fold gap between your desktop and your mind. This computer of the future will, of course, be very different from the Compaq or Macintosh you recently bought at Gadget City. Like your brain, it will operate in three dimensions (not two as computers do now with their flat circuit boards), and it will be doing 20,000,000,000,000,000 (20 million billion) calculations simultaneously.

Still, this won't be enough, Ray admits. These numbers and speeds would have stunned Turing himself, but they only address one part of the artificial intelligence equation. Horsepower. Processing speed. Cycles. Speed and power are important, says Ray, but they do not equal intelligence, no matter how many zeros are involved. You can rev up the fastest car on the planet, let it go at the Indianapolis 500, and all it's going to do is run into the nearest wall, unless you have a driver behind the wheel. The point is, a brain without the programming to run it, without the driver behind the wheel, is just a big fat engine with nowhere to go.

Or put another way, capacity is one thing, real intelligence is another.

So how do we manage that?

Shall the clay say to him that fashioneth it,
"What makest thou?"

—Isaiah 45:9

14

RAISING CONSCIOUSNESS

Let's assume that Moore's Law holds, and Ray Kurzweil's prediction is correct. In twenty to thirty years we are swimming in outrageously powerful thousand-dollar computers capable of human-level thinking. Once that's the case, the question then becomes: How, exactly, do you manage to create life-forms that exhibit human level intelligence but happen not to be human? (Never mind for now *why* we would do this. That comes later.)

Truthfully, no one has a clue (not even the most creative *Star Trek* writers). Even if we had all of the processing power in the universe, nobody knows what intelligence is, let alone how it works or where it comes from. The issue is wildly debated in university hallways and scientific and philosophical journals all over the world and has been for centuries. What is consciousness? What is intelligence? Can it be re-created or is there some impossible-to-explain, uniquely human quality that ensures that only we are capable of it? Is there a soul? And if so does it have to dress formally before going to heaven? Plato, Aristotle, Spinoza, Freud, Sartre, Woody Allen. The luminaries that have been flattened by this question would fill the bar on DS9.

And if you're looking to me for any answers, don't. They're beyond me in my personal life and I certainly won't try to tackle them between the pages of this book. All I can say is that based on my own experience, it's never, *ever* a good idea to say that *anything* is impossible.

Normally scientists might try to solve a problem like this by looking around for examples of conscious, intelligent life and seeing how *those* creatures manage it. But it turns out that the only species that everyone agrees is both conscious *and* intelligent (unless it has been in a singles bar for more than two hours) is . . . us. So the natural place to go and figure the problem out is inside of our own heads. The problem is: so far, the brain has eluded itself.

Or put another way, our own minds can't comprehend our own minds.

Reverse Engineering the Brain

Yet, says Dr. Kurzweil, there may be a way around this nagging problem. Don't try to take the brain apart and figure it out, just reproduce it.

"Someday, not very far down the road, we'll be able to develop the software of intelligence, the system that can put all of that computing power we've been talking about to use. In the future we'll be able to do a high-resolution scan of the brain. We already scan the brain—that's what MRIs (magnetic resonance imaging machines) do. We can't do it fast enough yet, or precisely enough, to get the really detailed scanning we need, but all of that is accelerating. Well within thirty years we'll be able to scan the human brain, see every circuit, and then make very precise copies."

Without really understanding how it works?

Kurzweil is thoughtful. "Well, I could see a couple of scenarios. On the one hand we could just re-create a computer version of the brain without necessarily understanding the whole global process, you know, how the brain really works. This means we make a digital copy of the brain in the same way a fax machine copies a document without really paying any attention to the actual content. You'd have to understand local brain processes, but not every detail. You would make an extremely high-fidelity, digital copy, and once that copy is made and operating, it would act like the original human. If it has been copied accurately enough, and in enough detail, I don't see any reason why that method wouldn't work. Anyhow, that's one approach.

"But the most interesting—and, I think, the most likely—scenario is that we will reverse-engineer the brain. We will figure out how it does everything that it does, but we'll do it backward by kind of taking the brain apart after we've gotten a really good look at it. Just as the Human Genome Project is making it possible to reverse-engineer DNA now. We have all this data about the human genetic code, we have it described right down to the last gene, but we still don't know what it all means. We know where all the genes are, but we still don't know what they do. It's going to take us ten to fifteen years to actually understand the human genome.

"We'll do the same thing with the brain. We'll get an extremely detailed description of it, a massive scan. Then we'll start to work backward to understand what the information means. We'll learn the secrets of how it processes information. It will give us insight into human nature, human intelligence, human dysfunction, and also allow us to build computers that have those kinds of intelligent capabilities."

In other words, Kurzweil is conceding that the brain itself can't get to the bottom of how it works any more than we could hope to understand how the *Enterprise*'s warp drive works by standing in the engine room and contemplating it. But if we had a detailed picture and description of the engines, if we could go in there and take them all apart and closely observe what each piece does, *then* we could figure them out. Or at the very least we could create a detailed copy that would work like the original.

It still sounds like a pretty steep order to me.

The Weighting Game

Ray agrees, but he says we already have some insights into how the brain makes sense of the world; how it records and recognizes patterns.

How?

"We are bombarded every moment by huge amounts of information, right? A million bits a second come through each of our ears. Our eyes feed us something like 100 million bits per second. And that's just *two* sources of information. We can't keep all of that around. We can't make sense of all of it. So we have to boil it down into a much smaller amount of information, but it has to be information that is still meaningful. How do we do that? We selectively *destroy* most of the information. And the information that is left is a meaningful pattern.

"Scientists have observed this in simpler animals, and they've seen it in humans when they use brain scanning technology like PET (positron emission tomography) scans. If the brain is presented with a stimulus, say a picture of something, and you look at a real-time brain scan of that activity—a representation of what is going on inside the brain at that moment—it initially reveals that it's recording wild, chaotic signals; all sorts of neurons start firing off. It just looks like fireworks. Then, after a fraction of a second, the chaotic signals settle down into a stable pattern. And that stable pattern represents the pattern recognition decision of the network. The neurons that continue to fire represent the brain remembering the information that is important; it's weighting every bit of input and keeping what it needs to make accurate sense of the information it is getting. The rest of it gets tossed."

"So it is literally remembering only the information necessary to build a picture of what is an accurate picture of reality. Anything else that isn't necessary is thrown away?"

"Right. It's like a big organization. It might be very chaotic when it first begins discussing an issue, but then after a while everyone settles down and reaches some common ground, an approach that works. They come up with a consensus."

Great, my brain operates like the United States Congress. Actually on some mornings I think it's operating more like the Russian Duma.

"Anyhow," says Kurzweil, "we can actually perform the same kind of pattern-recognition technique in computers. The technique is called neural networking. Neural networks mimic, in a very simple way, systems that work in the brain."

And when Ray said that, the systems in *my* brain reminded me that I had already seen a terrific example of neural network technology when Chip and I visited the NASA Ames Research Center not far from Palo Alto earlier in our journeys. It was on the same ill-fated expedition that included Xerox PARC. NASA, I remembered, was loaded with bushels of amazing, futuristic technologies, but the one that let me fly an F-15 fighter jet *and* taught me how neural nets work, that was the best.

Neural Nets and Fighter Jets

Our story begins high in the desert sky above Israel in 1987. American pilots were running routine missions flying their aircraft. One was flying an F-15 fighter, the other what the military calls an A-10 Warthog. We don't know why they were flying so close to one another, but they were and for some reason the A-10 suddenly did an inverted "jink"—a fighter avoidance maneuver. In a split second the A-10 came up under the F-15's right wing and tore it off. In another second the A-10 was gone, burst into a thousand pieces from the impact.

It should have been the end of the line for the F-15 too, but somehow, after desperately wrestling with the jet's controls, her pilot found one stable position that kept it airborne. It wasn't easy to hold. If the pilot deviated even one degree or failed to hold the jet's airspeed between 320 and 370 miles per hour, the F-15 would start to lose control. But somehow he managed to keep the hulking, side-winding monster steady in the air. Now all he had to do was land it . . . at 345 miles per hour!

The jet must have hit the ground with a tremendous thud as it hurtled down the runway. I've done a little flying myself, and I can tell you that controlling a one-winged fighter jet at this speed would have been like wrestling an alligator. To slow the thing down before it shot off the runway, the ground crew extended a cable for its tailhook. The F-15 hit it traveling at 350 miles an hour and tore the hook right out of its belly. But the cable cut the fighter's velocity to 230 miles per hour. At that speed the pilot was finally able to brake the nasty, twisted, fractured mess to a halt . . . and survive.

How this pilot managed to fly and land an F-15, literally on a wing and a prayer, is a mystery not even he can explain. He just did it. In a few split seconds his mind had somehow summoned every instinct and hormone and shred of knowledge it could, and then assembled it all into a series of actions that enabled him to right the jet and bring it down.

When Chuck Jorgensen, NASA Ames's chief scientist at its Neuro Engineering Laboratory, got wind of this story, he saw an opportunity. He didn't know how the pilot had managed to keep the jet in the air, and in some ways he didn't care. What he *did* care about was that an F-15 could actually fly with one wing! The pilot had proven it, even though most engineers would have sworn that it was impossible. Jorgensen figured if the plane could be flown with one wing, then perhaps a computerized system could be built that would basically accomplish the same thing that this single, amazing pilot had, *except it could do it every time, no matter who was at the rudder, automatically.*

That, he thought, would be good.

And that's where neural networks come in. "This wasn't the sort of thing that some programmer could write a few thousand lines of if-then code for," Jorgensen told me as we sat in his lab at NASA. "You know, code that says, *if* this happens *then* do that. No engineer or programmer could possibly predict all the variables at work when a machine this complex is flying at supersonic speeds, especially if it was severely damaged. Just too many ifs. There is wind shear, barometric pressure, air speed, gravity, momentum. Accidents can also 'daisy chain'—one accident can cause another and then more—each in increasingly unpredictable ways. The unknown or unpredictable forces would be endless, and beyond comprehension, literally unpredictable."

But what if the jet was programmed, basically, to survive, Jorgensen wondered. What if a system could be built in such a way that it was intelligent and could learn to reach the goal it had been given? What if the jet could program *itself* to stabilize itself and fly even if it were badly damaged?

The Bathtub Baby Test

A child psychologist once told me that if you ever watch a two-year-old playing in a bathtub, you're watching a learning machine programming itself. When kids are playing, it's not *all* about fun. They are constantly engaged in making mistakes, learning from those mistakes, and then logging the useful information away. They don't think about or reason out what is happening. They are just doing it, learning by trial-and-error. They splash the water and learn about heat and density. They fill a small cup and pour it into a bigger cup and learn about weight and volume, gravity and momentum.

Studies have shown that the brains of children at this age are burning twice the energy of an adult brain. It turns out that inside that little cranium, the brain is actually manufacturing millions of brand-new synaptic connections on the fly! Those connections represent the experience they store for later use. It is a way of the brain saying to itself, "Better lay down some wiring so you can record this, it may come in handy someday." I am told that unlike the brain of, say, a dung beetle, the neural networks inside of our heads are not hard-wired from birth. A beetle is preprogrammed from the first moment of its existence to behave in very specific ways. Usually those ways work well enough to ensure survival, and when they don't—*splat!* bug juice.

Humans, it seems, are more able to adapt to change because significant portions of our brains are *not* hard-wired from birth. Our wiring continues to evolve long after we come into the world, changing again and again, all the way into old age in fact. But since the brain can't afford to keep *all* of the connections it makes, those that are not consistently used, and reused, are eventually discarded. In other words, the only connections that remain are those that experience shows are necessary.

You use it, or you lose it.

As I understand it, assuming I understand it at all, this is basically a slower, physical example of the chaos-turned-to-pattern brain scanning images that Kurzweil mentioned scientists see when they scan brain activity. In Kurzweil's example, he was talking about millions of neurons firing off within a few seconds. In this case we're talking about actual, physical synaptic connections being made over periods of months and years.

Jorgensen told me that it is this synaptic-connection trick that neural net technology imitates, in a very rudimentary way. Unlike a beetle a neural net computer is not programmed in advance to do exactly as it is told. Instead the program is given a general goal. To reach that goal it is allowed to try lots and lots of different strategies until something works. That's the chaotic, fireworks part.

"When a strategy moves the program a step closer to its goal," explained Jorgensen, "then that connection is strengthened or weighted, numerically. It's the software version of making a stronger synaptic connection, and it is a way of letting the software remember and learn the things that help it reach its goal."

Something like the two-year-old in the bathtub, I suppose. If another strategy works, that connection is also strengthened and then another and another and so on.

It was this approach that Jorgensen and his team used with the F-15. Of course before they did even that, they had to create a "bathtub" that they could play in. F-15s cost several hundred million dollars a pop, and its human pilots are far more valuable than that. So it wouldn't do to rocket a few pilots into the stratosphere, shear off a wing, and see if the software worked. Those flyboys are crazy, but not *that* crazy.

No, for testing, Jorgensen needed to create a virtual environment, a "sky" where accidents could happen, but with impunity and at low cost. So his team brought in more computerized firepower and created a digital sky where the jet, and the software operating it, could play. Next they gave their cyberjet a simple goal: stay airborne, stabilize, fly right—even if you have only one wing, or no tail, or whatever.

"Lots of code was written which simulated the conditions under which an F-15 flies, including most of the environmental variables," Jorgensen told me. "The jet's neural net was outfitted with as many as thirty-two nodes, virtual neurons that can make more than a thousand connections, enough to learn what it needs to learn to gain aerodynamic stability. This is nothing like the capacity of the human brain, but good enough to work within the narrow behavioral scope of a jet fighter."

Captain Shatner

Now enter . . . me. I stand in the cluttered, computer-laden NASA laboratory run by Jorgensen and his team—lots of young, intelligent faces. The entire scenario is explained to me, and I am then asked to sit down in front of a computer monitor, and fly the virtual F-15. It's not the bridge of the *Enterprise*, but, hey, I don't stand on ceremony. So I sit and place my hand on a control rudder that is linked to the computer. There before me on the monitor appears the jet, cruising just as nice as pie high over a virtual desert. Very beautiful, we're just streaking through the sky at one thousand miles per hour when . . . *wham!* Off comes the F-15's right wing. Immediately it dips and starts losing control. The joystick shudders in my hand, but then in another second, before I can even do anything, the neural net program kicks in. It tries increasing speed, it tries tipping the jet to the left, then the right . . . it tries everything, and it's trying it so fast that I can't keep up, or even know that it's constantly, subtly learning.

The virtual environment, the "sky," wants to send the damaged thing into the ground like a thirty-two-ton dart, but the software that Jorgensen's team has written is firing away, making connections left and right in an effort to accomplish the goal it has been given—stay airborne! It rolls, tumbles, dips, and then, as if alive, it finds the correct combination of working strategies and . . . it flies.

What Jorgensen and his team have done is simulate a 68,000-pound, supersonic robot that can be threatened with annihilation and then "think" its way out of trouble. The F-15, at least this virtual version, was doing, in a very rudimentary way, what bathtub babies do. It was learning.

Very soon, Jorgensen's team realized you could also use this technique to design better planes, on the fly, so to speak, by creating digital F-15s with all sorts of different configurations. Put the tail rudder here, taper a wing there, then try it out, see if the jet can fly faster or snap roll better. It's the closest thing to letting a jet design itself so far (which is exactly where this technology is heading).

Machines that can actually learn in a way fundamentally similar to the way we do is a pretty powerful concept. Granted, thirty-two neurons isn't exactly simulating Albert Einstein, but if you can redesign an F-15 using this relatively simple approach, imagine what you could do with thousands or millions of virtual neurons.

Mixing Humans and Machines

And that brings us back to Dr. Kurzweil who wonders what would happen if you not only loaded up a computer with neural net software, but combined it with the raw, number crunching power in more standard computers? Could you create an intelligence that is smarter, better, faster than ours? Ray suspects we could.

Remember Deep Blue, the computer that beat Kasparov? The software that made Deep Blue so smart wasn't neural net or pattern recognition software. It was a programming strategy called the "recursive formula." The recursive formula, says Ray, is pretty much the programming behind the software that runs your desktop computer. It is also why your desktop is so stupid!

The problem with the recursive formula is that it's not very flexible. It's deep but thin, like a wedge. The whole success of the recursive formula, as Kurzweil puts it, depends upon "stating the problem precisely" to the computer. In other words telling the computer exactly what you want it to do so that it will never become confused. The program commands the computer and the computer does what it is told, precisely. In the case of a chess program, Ray explains, the computer's instructions would read something like this:

1. Your opponent just moved here.
2. Now calculate the best of all possible countermoves.
3. Also calculate your opponent's countermoves for each of your countermoves, and so on.

"Basically," explains Ray, "the computer is programmed to 'think' along these lines: 'If I go here, he might go here. If he does, then I'll go here. But if I do *that*, then he might go here. So then I'll go there.' It's actually doing 200 million of those kinds of move–countermove calculations a second."

We poor biological entities, on the other hand, have neither the computational horsepower nor the time for that. But if we can't do 200 million computations in a blink, then how do we play chess as well as we do?

You guessed it. Pattern recognition!

"A chess master like Gary Kasparov has studied chess, thoroughly," says Ray. "He's actually mastered a hundred thousand different situations. And when he's playing, he rapidly recognizes the situations he sees on the board as being similar to ones he's studied. And he tests what he is seeing now against what he knows; he compares the patterns and makes a move. He thinks,

'Oh, yeah. This is just like that move that Grand Master So-and-so made a year ago when he forgot to protect his bishop's pawn. I'd better remember to do that.' Or he may simply, unconsciously, recognize something odd or troublesome about the pattern of the pieces on the board. It's very subtle."

So, suggests Kurzweil, how about if we combine the brute number crunching force of the recursive technique with the neural net architecture that enables learning? "*Then* you might have yourself a machine that can learn, actually gather all kinds of knowledge, *and* do it very, very fast."

How about if we do something like that?

Then what?

Then you get on a jet and fly to Roboburgh.

When a distinguished but elderly scientist states that something is possible, he is almost certainly right. When he states that something is impossible, he is probably wrong.

—Arthur C. Clarke

Simple things should be simple and complex things should be possible.

—Alan Kay

15

DATA-PHILE

Roboburgh is what *The Wall Street Journal* calls Pittsburgh, Pennsylvania these days. Why? Because the city has shed its old steel town image and stepped up as one of the nation's leading high-tech centers. Among other advanced technologies, it has developed a particularly impressive expertise in robotics. In the past several years half a dozen robotics companies have set up shop in the city's warehouse districts and enterprise zones, and in 1996 it became the home of the National Robotics Engineering Consortium (NREC), a collaboration between Carnegie Mellon University, NASA, and several robotics start-ups.

Being the home of Carnegie Mellon University hasn't hurt this transformation. Along with Stanford and MIT, Carnegie Mellon is consistently rated as the nation's best in the field of computer science. And one of the reasons the school has reached such rarefied status is because it has a faculty full of wildly creative thinkers like Hans Moravec.

You remember Hans, the *other* rabble-rousing intellect I mentioned a couple of chapters back who, along with Ray

Kurzweil, is focused on what folks who run in these circles call the "trans-human condition."

That's why we winged our way to Roboburgh.

Robo Planet

I've been to a lot of "strange, new worlds" in my time, but most of them were fictional. I guarantee you that I never witnessed anything in real life quite like the things I saw as I walked the hallways outside of Hans Moravec's office at Carnegie Mellon's Robotics Institute. There were adolescent robots motoring along with their little camera eyes rolling as they tried to figure out the world. I saw multiple species of machines: some with faces, some trundling low to the ground, some big and bulky chattering along on tiny wheels, outfitted (I later learned) with arrays of sonar and infrared sensors. I looked into bays with robotic cars (one of them recently drove itself across the United States without any human help for ninety-eight percent of the trip; it had a problem with tollbooths); and rooms strewn with wires, cannibalized computers and robot heads looking blankly (or was it bleakly) at the parts all around that could make them whole again. (None, I noticed, resembled Brent Spiner.)

Hans Moravec would smile or chuckle as we passed these machines. To him they were more like children or grandchildren than wired-up bundles of gears and motors. After all, each was, in some sense, his progeny.

Hans himself is a big man with a broad, open face that breaks frequently into childlike grins when he discusses the future. Despite his linebacker frame, football was not his game as a kid. He was too busy devouring science fiction books and rigging up various, oddball contraptions. The son of an engineer and a mother who encouraged scientific exploration, he has been fascinated with robots for as long as he can remember. He built his first one with the help of his father when he was four out of a kind of wooden version of an erector set called Matador.

"It wasn't very fancy, but by turning little gears with a crank, I could make the robot move its arms and legs and head. I remember looking at this thing and thinking how it resembled a human, even though it was mechanical.

By ten he had managed to build a pretty good self-powered robot out of tin cans, a motor, lights, switches, and a few batteries.

"That made a big, big impression on me," he says. "It made me wonder, what if you just kept making these things more like a human; just kept pushing them farther along, one step at a time?"

That's pretty much what he's been up to ever since.

Hans grew up in my hometown, Montreal, but while I had moved on to Los Angeles and was hopping around on the *Star Trek* set making the universe safe for humanoids, he was getting on with the work that might someday replace them—taking top honors at high school science fairs with creations like light-seeking, robotic turtles. He then went on to Stanford where as a graduate student he helped design and build a machine called the "Stanford Cart," the first mobile robot capable of seeing and navigating through the world around it without being programmed ahead of time to do it. (However, it did take the Cart hours to cross a room. Computers were slower in those days.)

Hans doesn't see the world the way most of us do, which is exactly why it was essential to visit him. He is a founder of what has become the largest robotics lab in the country, and gets paid to think about how to someday create what we in the science fiction business call androids. No human—at least no human with better credentials—believes more in the future of robots than he does.

This isn't to say, however, that people are comfortable with the conclusions Hans draws and the thoughts he thinks.

For example, in his first book, *Mind Children*, Hans predicted the demise of the human race. Just a small thing. He wrote that robots would inevitably replace humans as the dominant intelligent species on Earth, and he made a very convincing, reasoned case for it. Just the laws of evolution at work, he said. Nevertheless people were shocked. Not because Hans had presented any breakthrough technology, but because a credible scientist came right out and said that humans would one day become obsolete.

In that same book Hans also imagined a time in the twenty-first century when we will be able to transfer a human mind into a robot, creating a kind of indestructible, digital hybrid of ourselves. One passage in particular really gave me the willies. But at the same time it was enthralling, like the story of Frankenstein. Read it and I think you'll see why.

> You've just been wheeled into the operating room. A robot brain surgeon is in attendance. By your side is a computer waiting to become a human equivalent, lacking only a program to run. Your skull, but not your brain, is anesthetized. You are fully conscious. The robot surgeon opens your brain case and places a hand on the brain's surface. This unusual hand bristles with microscopic machinery, and a cable connects it to the mobile computer at your side. Instruments in the hand scan the first few millimeters of brain surface. High-resolution magnetic resonance measurements build a three-dimensional chemical map, while arrays of magnetic and electric antennas collect signals that are rapidly unraveled to reveal, moment to moment, the pulses flashing among the neurons. These measurements, added to a comprehensive understanding of human neural architecture, allow the surgeon to write a program that models the behavior of the uppermost layer of the scanned brain tissue. This program is installed in a small portion of the waiting computer and activated. Measurements from the hand provide it with copies of

> the inputs that the original tissue is receiving. You and the surgeon check the accuracy of the simulation by comparing the signals it produces with the corresponding original ones. They flash by very fast, but any discrepancies are highlighted on a display screen. The surgeon fine-tunes the simulation until the correspondence is nearly perfect.
>
> The surgeon's hand sinks a fraction of a millimeter deeper into your brain, instantly compensating its measurements and signals for the changed position. The process is repeated for the next layer, and soon a second simulation resides in the computer, communicating with the first and with the remaining original brain tissue. Layer after layer the brain is simulated, then excavated. Eventually your skull is empty, and the surgeon's hand rests deep in your brainstem. Though you have not lost consciousness, or even your train of thought, your mind has been removed from the brain and transferred to a machine. In a final, disorienting step the surgeon lifts out his hand. Your suddenly abandoned body goes into spasms and dies. For a moment you experience only quiet and dark. Then, once again, you can open your eyes. Your perspective has shifted. The computer simulation has been disconnected from the cable leading to the surgeon's hand and reconnected to a shiny new body of the style, color, and material of your choice. Your metamorphosis is complete.

My reaction? Gulp!!!

Passages like these raised the hair on a lot of academic (and not so academic) necks and attracted considerable attention, from AI researchers to cognitive psychologists to philosophers. After reading it, everyone had an opinion, and not all of them were favorable.

"Some people became upset," says Hans.

Over the last few years, Hans says he's backed off some from his mind-transfer scenario. This isn't because he doubts it some day will be technically possible, but he suspects that combining a human mind (and all of that greedy, angry, primal baggage that can come with it) together with the power and speed of a super-intelligent robot might be a *just a touch* dangerous.

"A creature like that might make the neighbors a little nervous," he says.

Yeah.

Since Hans unabashedly believes that robots will eventually run the show, I had to ask him when he thought we would see the first really intelligent, sentient, Data-like creations.

"Oh, I think by middle of the twenty-first century," he said, as if it were the most obvious thing in the world.

Really! Data by 2050. Not even Gene Roddenberry would have predicted that! In fact Data doesn't show up until 2335. But Hans is not talking three hundred years down the road, he's looking at fifty! That's as near to us as Turing himself. That means that a lot of you reading this may see a world filled with a whole new species of creature: very advanced, homemade creatures that will not only shine your shoes, but discuss politics with you!

So what does it take to get there?

Well, for one thing a body.

Duh, Bill. *All* androids have bodies. That's what they are, machines made in our own image and likeness, right down to their arms and leg, heads and torsos!

Yes, yes, that's what I thought too, but it turns out we are missing the point. The way Hans sees it, an advanced robot doesn't have a body because it's supposed to *look* like us, it has it because a truly intelligent machine—at least one intelligent in the ways that we are—isn't possible without one. He says a box, even a highly perceptive box loaded with computing power, can't gain a true understanding of the world if it can't move through it, and sense it, and gain the knowledge that moving through and sensing the world delivers. In other words, no body, no human-level intelligence. Like the song says, "You can't have one without the other."

I Ain't Got No Body

This was fascinating to me because I don't usually connect high intelligence with arms and legs. But when you think about it, needing brawn in order to develop brains makes a lot of sense. After all, our bodies *are* great sensors, linked beautifully to both the external world around us and the internal one of our minds. Our bodies also deliver a lot of that common sense knowledge that Marvin Minsky says is so essential to human intelligence.

Hans points out that we accumulate common sense not so much by thinking about it, but by experiencing it, directly. We learn to avoid things that are too hot or cold, too loud or too bright. We feel the warmth of a hug and can sense the meaning of a kiss. Can't do that without a body. We absorb this information without reflecting on it, connect it intricately with other information and unconsciously develop . . . knowledge. In other words, we learn by experience, but how much could we possibly experience without our bodies?

Not much.

There's plenty of evidence that physical connections are essential to our humanness. Some studies, for example, show that infants who aren't held and coddled, and who live their early years without meaningful human contact, will develop severe behavioral and mental problems later in life. They often seem . . . well, machine-like, emotionless. It's as though the human part of their brain was never switched on.

Or suppose we could disconnect our brains somehow from our bodies and place them in a solution that kept them alive, literally disembodied? What would happen? Well, it turns out that researchers have placed subjects in sensory deprivation tanks, pools of water inside entirely enclosed, utterly dark and silent places. After several hours, many subjects begin to hallucinate. This isn't because they're bored or hungry or want to watch CNN; it's because the brain isn't getting enough contact with the world through the body it's attached to. So, after a while, it begins to manufacture its own reality. We do this when we sleep too. It's called dreaming.

So having a body is a big deal. The way Hans see it, a creature as intelligent—as conscious—as Data could never exist in the first place unless he was robotic (or human-like). Unlike Deep Blue, Data *would* come in out of the rain, but not simply because he has the legs to do it but because his legs (and eyes and arms and ears), would have helped him to develop the *intelligence* to do it.

However . . . creating mechanical bodies that work like ours has proven to be a monumental pain in the processor. The first lessons in this came way back in the 1950s, says Hans, when early computers were busy astonishing the world playing chess and solving mathematical theorems.

At the time, the early AI gurus figured they had already handled the "hard stuff"—you know, things like calculating and "reasoning"—so now all they had to do was polish off the easy, physical stuff. One of the first challenges they tackled was building a robot with "arms" and "eyes" that could play with blocks. Well, maybe "play" isn't precisely the right word because that would have meant that the machines were experiencing pleasure, and the truth is that they weren't experiencing anything.

Anyhow, playing with blocks seemed a good place to begin tackling the physical issues. After all, any toddler could play blocks, right? So they constructed machines programmed to identify, say, a red block, grab the block, and put it on top of the blue block. Piece of cake.

Except it wasn't. It was hard. Very hard.

First the machines couldn't figure out what a block was, period. Seems crazy, but think about it. How do you explain what a block is to something that has no experience of "blockness." The visual systems also had a hard time knowing where one block ended and another started, or for that matter where a block ended and *thin air* started.

Then there was the dexterity problem. It turned out to be monumentally difficult to create a machine that could pick up a block gently enough not to topple the entire stack. Color recognition was another issue. After all, how do you teach a machine what a color is? (Imagine trying to describe a sunset to someone who has been blind from birth.)

So it turned out that the "easy things" were not.

Scientists are still learning lessons like these, says Hans. Take, for example, the problem of making machines that can do two things that our bodies do all the time without thinking twice: seeing and walking.

Eyes: No Mean Feet

Creating working eyes is tough duty. Hans knows this because he has spent countless hours trying to figure out the problem. You can't compare eyes to cameras, he says. Cameras can record images, but that's nothing close to the seeing, moving, and intel-

This painting is the work of Aaron, a software cyberartist developed by artist and computer scientist Harold Cohen. Cohen taught Aaron to paint. The trick? Pattern recognition. Aaron's work has appeared in museums all over the world. Is there a creative side to AI?

ligent processing that *our* eyes routinely do. "Seeing is a whole other issue." After thirty years of toying with digital visual systems, for example, experts like Hans have only recently developed artificial eyes that can process an image that looks, at best, like a reasonably good watercolor painting.

This is amazingly good, for a robot.

Our own eyes, on the other hand, view the world in full color and high fidelity thanks to the many innovations that evolution has heaped upon us over the eons. The retina, says Hans, which senses the images around you, is about one quarter of the size of a postage stamp and way thinner, yet it consists of about 100 million neurons all interconnected to the optic nerve which runs through your brain to the visual cortex in the back of your head. Each of these neurons detects a million points of information—light, dark, vertical and horizontal edges, depth . . . you name it. Furthermore, he says, it gathers this information ten times per second so we can see everything from an attacking saber-tooth tiger to the smiling love-of-our-lives walking toward us from across the room.

Robots aren't there yet (especially the love part). They not only need tons of computer power to actually see a single static image, they need even more to do *the one thousand million calculations per second* necessary to look from left to right in real time!

Okay. That's one problem. (Hans is working on that.)

Now take walking. Well, you might say, how hard can this really be? One-year-olds walk for gods sakes. Wildebeests do it within a minute after they're out of the womb! *Bonk!* Out they come onto the Serengeti and they're up and running with the herd. If you are sitting in a chair reading this book, and decide you want to get a snack, you stand up and put one foot in front of the other and the last thing on your mind is how you actually manage to get from your chair to the refrigerator. But ask anyone who has tried to design a two-footed, walking robot (this is a rel-

After years of hard work, Hans Moravec now has his robots seeing this well in three dimensions. It's a huge leap, but it's not at the human level . . . yet.

atively small group) and he (or she) will tell you that the simple act of walking (never mind running, hopping, or skipping) is *the* mechanical engineering nightmare from hell.

Back in 1986, says Hans, scientists at Honda Motors, the giant Japanese auto maker, found this out the hard way. They decided to tackle a little thing they called the "Humanoid Robot" research and development project. The idea, according to Honda, was to create "a truly useful, intelligent, and highly mobile robot," one that would look and act a lot like us. A precursor to Data, you might say. Fifteen years later, more than $1 billion and, I'm guessing, a couple thousand gallons of Maalox, they finally managed to make some pretty impressive strides in the form of a five-foot-tall, two-hundred-eighty-five-pound robot that can actually walk (see facing illustration).* Not only that, he can actually kick a soccer ball. And he moves pretty smoothly and gracefully while he's at it. (If you'd like to take a look for yourself, check out http://www.honda.co.jp/english/technology/robot/concept1.html).

But it wasn't easy creating this walking machine, and the Honda team will be the first to tell you that P3's still a long way from human, even when it comes to putting one foot in front of the other. Why? Well, not that I understand this stuff, but here are just a few of the problems they say they had to solve:

1. Design and assemble multiple feet joints so that the robot could subtly shift its weight to avoid falling backward or forward or sideways. You think just standing around is easy, right? What could be simpler? It is for us, but not for a robot. Try standing still. Now close your eyes. Can you feel all of the small muscles

*Recently Honda completed work on Asimo, a smaller version of P3 that is just under three feet tall and one hundred pounds. They constructed a smaller version so that it wouldn't intimidate the humans it might someday serve. See the same Web site for more.

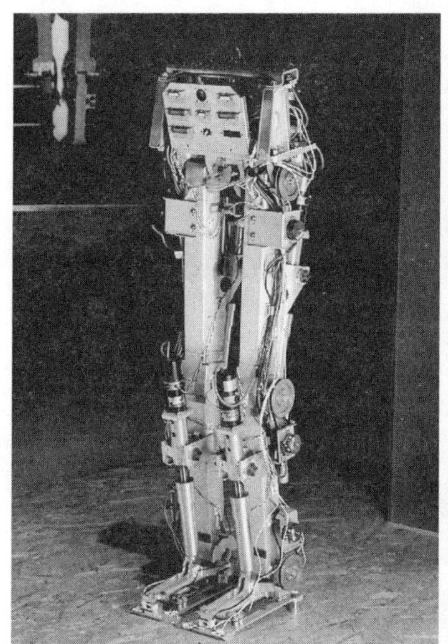

Asimo (right), a direct descendant of Honda's P3 (left), is now out there in the real world. He moves with amazingly human grace, but he wasn't easy to design and build, and he wasn't cheap. *(Honda Corporation)*

and tendons that are constantly working to keep you upright? That's what the team had to reconstruct. This helped solve the "center of gravity" issue so P3 wouldn't just topple over every time he stopped walking like a drunk walking the sidewalks of the Vegas strip.

2. Develop a "sense" of balance. The robot couldn't compensate and keep from falling over if it couldn't sense when it was in trouble in the first place, right? Humans have something called statoliths, three semicircular canals within the ear that detect the speed and direction of our bodies, or even different parts of our bodies in relation to one another. We also detect these sorts of things through nerves in our muscles and skin. Of course, our eyes and other senses also help make certain we don't constantly fall-down-go-

boom. To accomplish all of this the humanoid robot was outfitted with "one G-force and six-axial force" sensors. I have no idea what that means but the result was that these gadgets helped the robot detect what its legs and feet were "feeling" while walking. The researchers also had to outfit the little fella with an "inclinometer" as well as several "joint-angle sensors" designed to detect its overall posture. They did all of this just so P3 could maintain his balance, something we never think twice about (unless you've misbehaved at the annual Christmas party). Lesson: Upright perambulation is hard to do.

3. Finally, the project's engineers needed to develop a complex system of shock absorbers for the robot's feet so that (a) it knew when its feet were actually hitting the ground (this is handy knowledge), and (b) it wouldn't stomp around like a bad B-movie Frankenstein. In other words they wanted to make him light on his feet. We humans ease the impact of walking with all sorts of natural structures—fleshy skin, muscles, ankles and arches, tendons and joints. Without them we'd all walk like Bluto stalking Popeye. But addressing this problem is more than a matter of grace. The forces that our feet withstand when we walk are considerable, slamming down the equivalent of 150 to 400 percent of our body weight onto our legs and ankles and arches, depending on how fast we're moving. If we didn't have the right shock absorbing system, our legs and feet would shatter under the force. Same with the robot.

Those are just three problems the Honda team had to address so P3 could walk. Never mind the mechanics of the knees or the hips. Power and battery problems. Microchip and flexibility issues.

No wonder it cost a billion dollars!

Given all of these withering, mind-squashing problems, you have to wonder how Hans Moravec, or anyone else, can believe that creatures as complex as ersatz humans can possibly enter the world within the next five decades. If Hans is right that bodies capable of navigation and sight are necessary before we can hope to create sentient, Data-like androids, well, seems to me, in my infinite ignorance, that those problems need to be solved pretty soon.

Hans, however, is not terribly worried. He goes back to the Tribble Effect. Progress on all fronts, he says, is accelerating at an exponential rate. Metals and plastics and other materials are getting lighter and more flexible. We continue to cram more and more computing power into smaller and smaller spaces. Says Hans, "Advances in computer technology, medicine, nanotechnology, and materials engineering will produce universal robots that have not only developed 'human perceptual and motor abilities' but 'superior reasoning powers.' "

Okay. But still, I have to ask, How?

Hans has an answer. It's called evolution.

1. A robot may not injure a human being, or, through inaction, allow a human being to come to harm.
2. A robot must obey the orders given it by human beings except where such orders would conflict with the First Law.
3. A robot must protect its own existence as long as such protection does not conflict with the First or Second Law.

—Isaac Asimov's three laws of robotics; I, Robot

To you a robot is a robot . . . But you haven't worked with them . . . They're a cleaner, healthier breed than we are.

Isaac Asimov; I, Robot

I know that you and Frank were planning to disconnect me, and I am afraid that I cannot allow that to happen.

—HAL 9000 to Dave
2001, A Space Odyssey

16

YIKES! WHO'S IN CHARGE?

Evolution is time consuming. That's why you don't see much of it in science fiction. In a film or a book, you have to keep the story *racing* along. Evolution, even the robotic kind, doesn't race, it crawls. That's why we in the entertainment business don't burden ourselves with *how* you create robots or androids or cyborgs, they just show up, fully formed like fast food.

But Hans Moravec isn't in the entertainment business so he doesn't believe that someone can simply go away and think the whole robot problem through in advance, invent cool materials and more powerful computers, and then one day, like Victor Frankenstein (or Noonien Soong, Data's creator, for that matter), unleash a fully fashioned new life-form upon the universe. It's just too hard.

Instead he believes that robots will have to attain their human level intelligence the old-fashioned way. They'll have to earn it, one evolutionary step at a time.

"Nowhere in nature have we seen top-down programming where a species is engineered to handle certain things *before* it knows what those things are," says Hans, brushing his large hand back and forth over his bristle-short blond hair. "It's too difficult to predict in advance what might work and what might not. Besides, evolution doesn't consciously plan because it *isn't* conscious. Instead it programs from the bottom up.

"That means the most practical route to achieving the kind of behavior in machines that we humans exhibit is to recapitulate our own evolution," he says. "Put them out in the world. Let them struggle with it all on the most basic levels. When they fail, fix the problems and then send the next generation back out. Of course the actual machinery that will be inside them will be very different from ours, but in the end, I think we can get the same result."

The result, however, will *not* be like *Star Trek* in at least one particularly notable way. There are way more androids running around in Hans Moravec's future than in *Star Trek*'s. Like Frankenstein, Data was kind of a one-off, an aberration. In fact, when you think about it, *Star Trek* doesn't have much room in Starfleet for androids at all. Neither the shows nor the movies feature platoons of robots and cyborgs as part of everyday futuristic life, except as an occasional alien menace. (Could be a discrimination case here.) Humans are still front and center, running the operations, and Data is, no offense, a freak of nature, or technology, or something. Whatever he is, he's rare.

Again all of this makes plenty of sense if you're producing a science fiction series. It's easier to focus on one very interesting android, than a herd of faceless ones. But as I say, Hans doesn't see it working out like this in reality, which, frankly, I find damned disconcerting. His view of things is more Darwinian. He figures that the emergence of robots superior to us is simply the next big step not simply in the evolution of machines, but the evolution of life on earth. And when you talk about evolution,

DATA IN FOUR EASY STEPS OR HANS MORAVEC'S HISTORY OF THE FUTURE

UNIVERSAL ROBOTS—THE FIRST GENERATION
ETA: 2010
DISTINGUISHING FEATURES: GENERAL-PURPOSE PERCEPTION, MANIPULATION, AND MOBILITY

The first generation of universal robots won't be anything spectacular by Hollywood standards, just a machine with the processing capacity of a lizard and the personality, in Hans's words, of a washing machine. But it will be several steps beyond where we are now. Its primary goal will be to do exactly the kind of work we all hate. Pull the weeds, clean the house, assemble the stuff where the directions say "some assembly required" (now *that* will be a true test of artificial intelligence). They'll mostly run on wheels because wheels are simple and don't use much power. If they need to get up and down steps (not a trivial issue), they'll have special three-wheel assemblies or tracks that will get the job done. First generation robots will have a pretty keen sense of sight and get around the house or yard or warehouse without destroying everything in their path. (The exception, of course, will be my house where everything will go wrong. They'll trash the entire place and have a firefight with the talking light system for electronic supremacy.)

You wouldn't want "Gen-1" robots performing brain surgery on you, but their hands—probably more than two—will handle a dish well enough that it can put it in the dishwasher without smashing it to pieces. And when it needs a boost, it'll go plug itself into a wall outlet and juice up.

UNIVERSAL ROBOTS—THE SECOND GENERATION
ETA: 2020
DISTINGUISHING FEATURES:
LEARNING AND SIMPLE PROBLEM SOLVING

Second generation robots appear ten years after the first. Moore's Law has been hard at work, and IQs will have increased considerably, says Hans. Second generation robots will have the aggregated brainpower of a mouse, thirty times more powerful than its lizard-headed predecessor. Run a mouse through a maze and after enough mistakes, it figures out where the cheese is. Not exactly Albert Einstein figuring out the laws of the universe, but it's learning. Second generation robots will be able to do the same, possibly using neural nets, or programs called genetic algorithms, which are basically software versions of Darwin's "survival of the fittest." In this case, survival of the fittest knowledge and memories. Behavior that works, says Hans, is rewarded (strengthened internally); behavior that fails is weakened or eventually discarded.

Whatever the case, this generation of working machines will do most of this learning by making mistakes and correcting them on their own. "Oops, that's not a shoe, that's Tammy's doll! Tammy's doll goes *here.* I better go find the other shoe." If the robot is having a really hard time figuring things out, it might simply go to its owner and ask, "What, in the name of Asimov, is this?" Or some such phrase. This, at least, is better than what your standard mouse does.

You will also be able to train your Gen-2 robot. You could walk one through the daily cleaning routine, show it where things belong, and then let the neural net do the rest. It'll also teach itself to do things more efficiently, just as many people who are much more organized than I am often do in real life. This is a huge advantage over the first generation robots because it means Gen-2s will solve most of their own problems, basically tweaking its own software (unlike most teenagers).

To give the robots even more intelligence, Hans suggests that prototypes could be plugged into supercomputer simulations of real-life again and again before being cut loose in the real world. This way they can make lots of mistakes breaking virtual china rather than the real thing. Naturally, advances in materials research, power systems, and microelectronics (this means they'll be smaller and stronger) will enable Gen-2 to pretty much navigate anywhere most humans can.

UNIVERSAL ROBOTS—THE THIRD GENERATION
ETA: 2030
DISTINGUISHING FEATURE: WORLD MODELING

Reminder: everything is improving at an exponential rate. Gen-3 robots now have the processing power of a monkey and they are performing three trillion instructions per second, way smart for a machine. Basically what we have now is an extremely mobile supercomputer. That means that rather than making mistakes in the physical world and then adjusting for them, they will think ahead; work problems out in their "minds," Somethng like we do. When you combine this kind of brain power with more powerful robotic senses (hearing, smell, sight), Gen-3s will basically be capable of modeling the world around them in high fidelity, moment by moment, at least as well as your average howler monkey. Because its processing speed will be so great, Gen-3s will not only be able to watch us do things and learn the job, they'll be able to figure out ways to do the job better and faster. They'll even be able to share what they learn with other robots via wireless communication, a kind of robotic telepathy. Of course their personalities will also have considerably improved, though they aren't likely to have yet achieved the charm and savoir faire of Captain Kirk. But maybe with some work they will achieve Spock-ness.

UNIVERSAL ROBOTS—THE FOURTH GENERATIONETA: 2040
DISTINGUISHING FEATURE: REASONING

Now it gets scary. With extremely small and outrageously powerful computers processing information at the same speed as the human mind (100,000,000,000,000—100 trillion—instructions per second), Moravec predicts Gen-4 robots will not only be able to navigate, see, and sense as well or better than we do, they will be conscious and fully capable of reasoning. Humans may still debate whether these machines are conscious, he says, but the strange thing is that we'll be debating the issue with the *machines* as well as one another. *They* will think they are conscious, Hans points out, and they will be perfectly capable of having long conversations explaining why. Come to think of it, given the processing power at their disposal, it might be better to avoid a debate. You'll probably lose.

It's at this point that all of the learning from the previous three generations of robots will meld with the high-end, reverse-engineered programs that Kurzweil talks about to create a superintelligent being. These androids will pretty much know and understand everything that goes on around them. Even the behavior of Regis Philbin.

Handling the physical world will only be a small part of the issues these creatures deal with because they'll have a very rich mental, some might even say psychological, life. They'll sense when a human is angry or sad or happy, and react accordingly with fear or sympathy or joy (never with aggression, right, Hans . . . Hans?).

They are going to be awfully smart. So smart they may become a little obtuse. However, being sympathetic creatures, if they get the impression that you don't understand what they are saying (your crossed eyes might provide a hint), then their internal programming will tweak itself and they'll make an effort to speak at your lowly level. (Just as many scientists did for me and Chip when we researched this book.)

What will be the difference between a Gen-3 and a Gen-4? Hans put it this way: "If you ask a third-generation robot 'Why did you put flowers on the table?' it would answer, 'Because I thought it would cheer Hannah up.' And if you ask 'Why would you want to cheer Hannah up?' it would say, 'Because I like it when she is happy.' " And that would be the end of it. Like a child, Gen-3 wouldn't really be able to explain *why* it likes for Hannah to be happy. It just does.

"On the other hand, a Gen-4 could reason and explain why, in detail. "Well, I know she and Dexter have been having problems lately. It's been frustrating for her; he doesn't *ever* seem to listen. Sometimes he gets self-absorbed, don't you think, and it's like he's gone off to Altair IV. This upsets Hannah a lot, and . . ." Well, you get the idea, it would have plenty of feelings it could express.

"Beasts," said the British philosopher John Locke, "abstract not." Maybe, but Gen-4 androids will abstract, a lot. "They'll be powerful thinkers," says Hans, "able to draw conclusions from subtle observances. Unencumbered by the baggage of human evolution, they'll be able to apply concrete simulations to very complex ideas."

Imagine that. (*They* will.)

you also have to talk about extinction. That's just the way it works.

"An animal or plant or insect changes because of random mutations in its DNA," says Hans. "If the change makes the organism more successful, it survives and passes its genes on and you get more creatures like it. If not . . . well, say hello to good-bye.

"Anyhow, the harsh laws of survival hone the programming and have managed to double the size of its biggest brains every fifteen million years, on average. That's not exactly a blistering pace from our point of view. We can do better than that with robots because the improvements we create aren't as random as they are in nature. We can consciously look at what works, and what doesn't, and then make the appropriate changes. If we see a robot is having a hard time recognizing where a ledge is, for example, we can program it to avoid going over ledges and destroying itself. We don't have to wait for it to actually do it first."

So through this hybrid approach that on the one hand helps robots learn faster than their natural counterparts, and on the other, puts them out there in the real world to learn on their own, Hans figures you can move the evolutionary process along at a pretty good clip that will create smarter robots sooner.

Not everyone thinks this way in the world of robotics. A lot of approaches have avoided forcing robots to deal with the unpredictability of "life" by conforming the world to them. (Reminds me of some parents I know.) For example, some mobile robots are provided with a complicated map of a specific building they work in. They can navigate that building pretty well as long as nothing changes or deviates from the map they have had dumped into their robot heads. Problem is, like Deep Blue, they're idiot savants. They do one thing, and one thing only, well. Put them in a different building, and they're as lost as a two-year-old in a bad mall (or even a good mall, come to think of it, if there is such a thing as a good mall).

Another approach is to build robots that follow electronic "scents" like some high-tech insect. Again, that works fine, as long as the tracking system is there and nothing changes. Put the robot in another place and it's as useless as a Vulcan at an orgy.

To Moravec, and the others who work in Carnegie Mellon's mobile robotics lab, these are fine short-term approaches, but if you want to design what he calls a universal robot—one that can *really* get along in the world without a lot of attention and maintenance—you need to make it adaptable. In short, it has to be able to learn. This is tougher in the short term, but makes a helluva lot of sense if you want to avoid a world full of mechanical idiot savants. I mean don't we have enough of those already?

Not that this approach doesn't have its occasional downside. Take the case of Xavier, the Muscle-Bound Robot. I learned about Xavier when we visited Sebastian Thrun and his team at Carnegie Mellon's Robot Learning Lab. Sebastian codirects the lab, which specializes in a kind of robotic version of job training, or, if you think about it, robot rearing. Sometimes this backfires.

Xavier, an early Carnegie Mellon mobile robot. He blew the doors off the robotic world, so to speak. Not only could he deliver the mail, he could tell jokes too. *(Carnegie Mellon Robotics Institute)*

It did with Xavier, at any rate, a robot shaped like a barrel, four feet tall and a couple hundred pounds. He runs on wheels and can go backward and forward and pirouette on a dime. He was named after Xavier, the leader of the comic book team the X-Men, because he's smart, but physically handicapped like his namesake.

Anyhow, just to make certain that Xavier wouldn't pirouette into anything he shouldn't, he was initially outfitted with an array of sensors that bounced infrared rays off the objects around him. This enabled him to "see," sort of like the way a bat sees in the night by using a natural form of radar.

On the day of Xavier's maiden voyage he stood in the robotics lab ready to roll. Now the section of the lab where Xavier lived is a big concrete room with two huge, heavy wooden doors, each a good twelve feet high. (I saw them. They're *big.*) The doors exit into a university hallway where faculty have offices and where students bustle from classroom to classroom and lab to lab. Xavier's job was to roll into the hallway, navigate his way around the people and objects he encountered, and deliver mail to faculty offices, all on his own. If he ran up against an obstacle, his sensors were to let him know so he could stop, figure out a new route, and then move on with his appointed rounds. As an extra bonus, he also told knock-knock jokes.

"Knock-knock."

"Who's there?"

"Xavier."

"Xavier who?"

"Zave yer time. Let me run that errand."

"Zave yer breath, this is as funny as it gets."

"Zave yer self from these awful jokes, turn me off."

That sort of thing.

So the first day Xavier is turned on, he dutifully starts to move out. Problem is the big wooden doors haven't been opened yet. But that's okay because Xavier was outfitted with infrared sensors, meticulously designed to help him see obstacles, like closed doors. Except somehow his sensors didn't sense the doors. So he simply crashed right through them! He crashed through these two huge slabs of wood which weigh in close to half a ton, like they're balsa. This is not a robot you would want to cross. (Though you might want to sign him up for Wrestlemania.)

Luckily the stunned team was able to grab Xavier and quickly disable him by performing a digital lobotomy before he rolled through any more walls. They then guided him back into the lab for a little more tweaking, and eventually he went onto a successful career as a mail carrier. Though I hear his career as a comedian failed miserably.

This was back in the early nineties, ancient robotic history now. Since then the lab has made a lot of progress in robot evolution, says Sebastian. "We developed a next generation robot named Chips that now works full-time at the Carnegie Museum of Natural History here in Pittsburgh. He roams the dinosaur hall, giving lectures about various exhibits, complete with a screen that displays supporting video right from his midsection."

Apparently any day of the week you can see Chips, complete with bow tie and smiling face, wheeling around among the bones of dinosaurs that have been dead a couple of hundred million years, all with a retinue of gawking kids in tow.

Like Xavier's door mishap, Chips's experiences in the real world have turned out to be a great experiment in robot evolution. One of the first lessons the robot designers learned when they put Chips to work in the museum was that kids, at least some kids (ones like me), like to play "stump the robot" by standing in front of him and blocking his way. Now if your job is to help robots evolve to handle "life" as we know it in the twenty-first century, how do you deal with that?

Well, at first Chips was programmed to just navigate around the little ankle-biters. But a lot of kids just kept moving and blocking his way, so the creators modified him. Now when kids do this, Chips simply stops and waits, with the patience that only a machine can possess. After a while, the offending rug rat gets bored and wanders away, usually. If he doesn't, Chips whips out a phaser, annihilates the brat on the spot, and e-mails a notice to his parents.

Just kidding.

The researchers at CMU said they couldn't get the phasers working.

Chips has other impressive talents. When he runs low on power, he tells everyone within earshot that he's "tired" and rolls over to an AC outlet, plugs himself in, and has an energy snack. (Wish I could do that.)

Even with these nifty capabilities, Chips does not represent the highest state of robotic art, not anymore. No sooner had the Learning Lab gotten Chips working when they started designing a successor named Minerva.

Express Your Feelings

With Minerva, Thrun and his students really tackled the CMU lab's commitment to putting robots in challenging situations, no holds barred. Building on the experiences of Xavier and Chips, they cut Minerva loose as a guide in the Smithsonian Institution's Museum of American History in Washington, D.C. This is one of the busiest museums in the world, visited by millions from all over the globe each year. Minerva's job was to patiently answer any questions any visitor had and provide guidance and directions for them to the museum's many exhibits.

Every day, for two weeks at the height of the summer rush, Minerva faced thousands of pesky kids *and* adults who wanted to mess with her mind by standing in front of her, waving caps over her eyes, tossing things in her way. Occasionally they even asked for directions or for information about an exhibit. Though a far

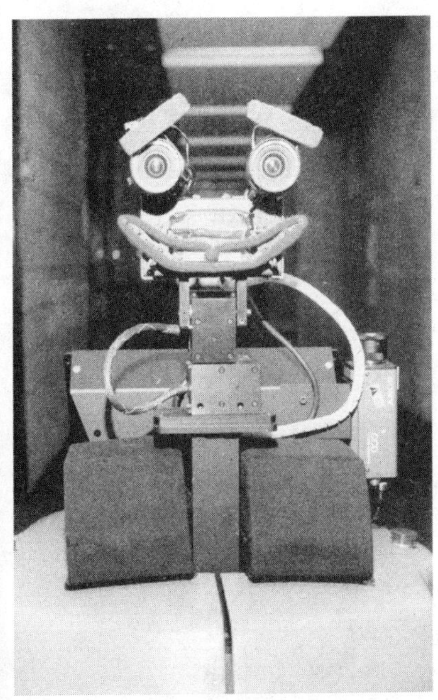

Minerva survived working at the Smithsonian in Washington, D.C., and interacted with thousands of tourists. Never once did she vaporize a human with a phaser. She *did* give them directions and answer their questions. *(Carnegie Mellon Robotics Institute)*

cry from Data, Minerva still showed her mettle. Never once did she vaporize a visitor.

She was able to handle the situation so well, says Sebastian, because she had learned from her siblings.

Chips, for example, uses a series of colored cards strategically placed around the Carnegie Museum's dinosaur hall to help him track his position. When he sees a particular card in a particular place, he understands where he is. The Smithsonian was too big for that approach; besides, the lab wanted to push the envelope.

"So we not only gave her a more powerful brain," says Sebastian, "but we added two stereoscopic, video camera eyes for the first time, and two sets of laser infrared range finders and two sets of sonar sensors. With all of these she really had a lot of navigational power. She could 'see' pretty well."

Once cut loose inside the museum, Minerva applied her memory, eyes, and sensors to creating her very own internal map of the entire floor, basically memorizing the museum's layout on her own. This is important, says Sebastian, because it means if anything moves later, she can redraw the map and handle the change, something very roughly similar to thinking.

Another big innovation was Minerva's looks; she got more of that than Chips, too. Not that Chips hadn't been given a handsome face, for a robot, but it wasn't very . . . expressive.

Minerva's, however, is animated, and that gives her more personality. From the neck down she looks like a steam cabinet on wheels—no legs, no arms, no navel—but from the neck up she's more human. Well, actually she's more like a caricature of a human with big cartoon camera eyes, huge blue eyebrows, and a big red mouth. But it works because these features can express emotion, which makes her more effective with the humanoids she deals with.

Normally Minerva's red mouth smiles and her eyebrows are raised at a friendly angle, you know, inside ends up and outside ends down. That's her happy face. But Carnegie Mellon's lab also programmed in a neutral face, a sad face, and a don't-cross-me-or-I'll-kick-your-ass face. If someone gets in her way, she's less passive than Chips. First, she gives you the neutral expression and says, in a pleasant, but no-nonsense female voice, "Could you please stand behind me?" Most people will step aside on that one. However, if, after a few moments you don't move, she gives you the nasty look—eyebrows down, eyes glaring—and she says in a voice that instantly reminds you of your mother when she had reached the end of her rope, "I need to get through!"

You should see the kids scatter on *that* one. Amazing how something so simple, yet so primal, works.

Interacting with Minerva is a hoot. You can't talk with her, but you can get information by touching a menu of questions on her video screen chest. (Try *that* with a human tour guide.) Instantly she provides an answer, or an associated tour. And says Sebastian, the carbon-based life-forms from other sectors of the country loved it.

There's one last Minerva ability I have to point out, her tirelessness. While at the Smithsonian, she worked twenty-four hours a day, seven days a week, and she was endlessly patient. When she wasn't fielding thousands of questions from tourists during the day, she was giving cybervisitors virtual tours at night over the Internet! After the museum closed, Japan and Australia were just waking up. Visitors could go online, visit with Minerva, and guide her around the museum, asking questions as she went along. Not quite as good as being there, but it illustrates an advantage robots have over us. They don't get sleepy or cranky and they take orders better than we do.

Now Carnegie Mellon's robotics lab is working on even more advanced robots, continuing to nudge them up their evolutionary path. The first of these is Flo, the Nursebot. (That's Flo for Florence Nightingale.) And her new cybersister Pearl. Pearl has been created since Chip and I visited the lab, but I did have a chance to meet an early version of Flo. She was naked, showing her hard drive and wireless modem and other electronic gizzards, but she didn't blush. Like Minerva, she has a flat panel screen under her chin, and articulated eyes, mouth, and eyebrows. Unlike Minerva, you can talk to her, and she can hear you.

Flo is designed for an especially tough market—the elderly. Today there are about twenty-four million people in the United States over sixty-five. By the year 2030 that number will likely double. But the real kicker is that the percentage of people who survive to over age eighty-five has already increased by forty percent just since 1990! That's a lot of old codgers.

So it occurred to some of the learning lab's graduate students that a dutiful, tireless servant available twenty-four hours a day might help extend the time that older folks can live a fulfilling, independent life. And that's Flo's job (and Pearl's too): help older people live better, longer.

Insane, right? Robots and the elderly? Bad combination. After all, elderly are set in their ways, and they tend to be intimidated by technology. Truth is any of us might feel threatened by a machine running around in our house. But then who

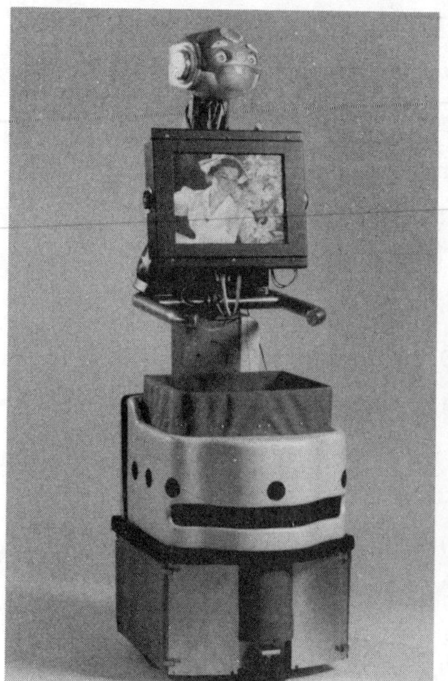

A close-up of Pearl (left), and interacting with residents (right), of the Elderly Care Facility at Longwood in Oakmont, Pennsylvania. She seems to be getting along pretty well. *(Carnegie Mellon Robotics Institute)*

could use a helping hand more than an elderly person who's not moving quite as nimbly as they used to, and what better way to prove that robots can be a positive, friendly force in society than to build corps of them to help grandma and grandpa?

Despite my deplorable experiences with high technology, I'd be willing to experiment with having Flo around, and not because I'm getting long in the tooth. She's a pretty impressive machine. When I said hello to her, she smiled, actually smiled, and said hello back in a pleasant, female voice. I was told that no matter how many times I ask her where my keys are, she would answer patiently, and never say once to me, "Check your pockets, for gods sakes! Honestly, you couldn't find your ass with both hands!"

Patience, even in a robot, is a virtue.

I asked her what was on television, she quickly dialed up the Internet, downloaded what was on the major networks through her wireless modem and told me, at the same time she displayed it all on her little computer screen.

Sebastian tells me that Flo never made it out of the lab, but Pearl, her next-generation sibling, did. Pearl just completed an introductory tour of duty at the Longwood Elderly Care Facility in a Pittsburgh suburb where she was a big hit. Not one resident passed out from fright, and she didn't run over a single set of dentures.

The Nursebot Project keeps evolving its progeny, just like Hans says it should. Soon the lab hopes to create Nursebots that will be able to answer the phone, and, then, by using caller ID, tell you who's on the line. "Do you want to talk with your son?" she'll politely ask. ("Does he want money?")

Already this is considerably more sophisticated than your standard answering machine.

Future Nursebots will also be able to display pictures of the grandchildren on her screen, especially if one of them happens to call. This would be nice while talking with a six-year-old granddaughter. Eventually, video phoning will be possible.

Nursebots will also be designed to retrieve things for you so you don't have to get up and down out of the Barcalounger more than necessary. In an emergency, she'll even call the hospital or doctor. All you have to do is tell her, "I've fallen down and I can't get up!" just like in the old commercials. She'll also call the doctor if she doesn't see you around, or notices that you aren't moving. (Let's not go there.)

On the one hand, some of this seems goofy. You wonder, we put a man on the moon thirty-three years ago but this is the best robot we can come up with in the new millennium? On the other hand, these robots are impressive because they are adapting to the real world, and doing it in many ways, not unlike we do. That fits Moravec's view of the way cyberevents will unfold in the coming years. Each of the robots at the lab has learned from its siblings. Each is smarter and more human-like than the one before. If Hans is right, in coming years, each "species" of robot will push the envelope a little farther until we end up with some very smart machines running around.

The eventual result? Data. Actually, a whole race of Datas.

Planet of the Apes

"It won't happen all at once, but compared to other evolutionary changes it will happen pretty fast," says Hans. And it's not that they [the robots] will necessarily wipe us out. It not going to be like *Terminator*. In fact they may take very good care of us. You know, fix Earth up so that it's a wonderful place to live, handle all of our needs, and then go off and run the universe."

Hmmm. A race of Nursebots.

So we'll be left behind like a planet full of lowland gorillas (without the poachers) with all of Earth as our very own Rwanda. Why do I not find this comforting? A Garden of Eden as home, no need for work, personal fortunes made and stashed away by the androids, and smart machines we've created to handle everything, from walking the dog to running mega-corporations. On the surface it sounds good. Yet it bothers me. Why?

I think it's the being left behind part I don't like. I don't like the idea of having options like that removed. That's what freedom is about. Having choices and being able to make your own decisions. I think history, and several *Star Trek* episodes, bear me out on this one.

Unlike Moravec, Ray Kurzweil doesn't see us being replaced by our electronic progeny in quite this way. Instead, he sees a kind of metamorphosis. Digital technologies will become increasingly embedded (literally) in us. We'll enhance our senses, minds, and capabilities with digital prostheses that will make us increasingly intelligent and capable. We will, he speculates, literally implant ourselves with more capacity, greater health, longer life. At the same time, machines will become increasingly human, and when all is said and done, by the end of the next century, you won't be able to tell the difference between them and us. We will have become one another, both with a common branch on the evolutionary tree—*Homo sapiens* (may they rest in peace). The new species? *Homo roboticus* or *Homo neosapiens*—take your pick.

"It's not so much that we're going to have this alien invasion of intelligent machines coming over the horizon to compete with us," says Ray. " It's really already emerging from within our human machine civilization. And we're very much going to merge with that technology. We're going to use it to enhance human intelligence."

I think I like Ray's scenario just a touch better, but I'm not sure if it's because Ray is actually right or because I just find it more comforting that we all won't end up like a bunch of Neanderthal bones—ossified and extinct.

Either way, I'm concerned that a few decades down the road there will come this day when I command my personal android (with whom I will, of course, enjoy a wonderful friendship) to get me something, I don't know what, maybe a second scoop of Häagen Dazs Double Dutch Chocolate, and he/she will say, "Why?"

"What do you mean, 'Why?' I'll say. 'Because I'm . . . hungry.' "

And then my android companion will look me in the eye, very seriously, and say, "I'm afraid I can't let you have another bowl of ice cream, Bill."

Then, I ask you, who is in charge? *Then* who is the dominant species?

Maybe resistance *is* futile. Or maybe we're looking at the dawn of a new era when we will be free to be more human than ever, living in some Eden that leads us to embark on cross-galactic expeditions in search of strange new worlds. There could be advantages to that.

Either way, though, I vote we move cautiously before kicking things into warp drive.

PART FOUR

PLAYING GOD

Then God said, "Let there be light," and there was light. And God saw that the light was good, and there was evening and morning the first day.

—Exodus 1:3-4

Needed: A prime directive for the twenty-first century. The dangers and responsibilities of messing with the recipe of life, the postponement of death, shape-shifting matter, and the creation of alternate realities. Can we build the holodeck, live forever, and avoid paying taxes too? A short course on freezing your way to everlasting life.

Power tends to corrupt and absolute power corrupts absolutely.

—Lord John Acton

17

DIVINITY SCHOOL

Pop Quiz:

General Order Number One is . . .

A. What Patrick Stewart always had for lunch on the set of *Star Trek: The Next Generation*.

B. Under the laws of the United Federation of Planets, the right to bear phasers.

C. The Prime Directive.

D. Anything the IRS wants it to mean.

If you guessed C, give yourself a pat on the back and a Vulcan mind-meld. Any *Star Trek* fan worth his salt knows that the Prime Directive basically says that if you work for Starfleet and show up at a planet that hasn't yet developed warp drive, you had better stay hidden and mind your own business.

The idea behind the Prime Directive was that if you've got warp drive and the civilization you've come into contact with doesn't, you are very likely *way* more advanced than they are (at least technologically). So you have to keep your phasers and transporters and tricorders to yourself, otherwise it might be tempting to use them to manipulate the poor, benighted creatures and do pretty much whatever you like.

More to the point, the Prime Directive is designed to keep Starfleet captains like Kirk from playing God. After all, to paraphrase Arthur C. Clarke, any sufficiently advanced technology appears to be supernatural, especially to those who have never seen it before. And when armed with this sort of power, it could be tempting to play Zeus and toss a lightning bolt or two.

Now thanks to this book, I've been lucky enough to peek behind the curtain of the future and see some pretty hair-raising inventions. It occurs to me that we might soon need to invoke a kind of Prime Directive for ourselves as we plow on into the twenty-first century. When you have a chance to consider what lies in the chapters ahead, I think you'll see that we are already well on our way toward dabbling in "God" territory.

Artificial intelligence is a perfect example. If we successfully create intelligent machines—and let's not even quibble about whether they are really conscious in the way we are—have we only created an extremely complex computer, or have we created a new life-form? Do life-forms have to be biological? I'm not sure where you draw the line. Let's say that down the road your Sony robot has served you well for five or ten years. Could you simply chuck it into the trash compactor one day—basically killing it? My guess is that it would be a pretty devastating loss. At least as tough as putting Muffy the cat to sleep because she's lost track of where the litter box is. Probably worse. (I mean at least the robot isn't soiling the carpet.)

Godly Territory

The way I see it, we begin to get into godly territory when we start reshaping the course of natural events in powerful, conscious ways. Designing and building intelligent machines, for example. Or engineering tailor-made human life, something we will soon be able to do thanks to successful efforts to map the human genome.

Then there are entirely new technologies on the horizon, branches of science that have not yet born fruit, but will soon. Nanotechnology, for example, will enable us to create computers and moving machines no larger than a molecule. When you are operating at this scale, I am told, there is very little you can't change or create because this is the scale at which the building blocks of *all* matter, living and otherwise, is created. A nanometer is one billionth of a meter, an invisible size, a size at which things happen very rapidly at a very small scale; the size at which DNA and enzyme reactions take place; the size at which neurons and skin cells are built. In short, the scale at which the miracle of life unfolds. And like life, nanomachines can be programmed to replicate by herding up other atoms and molecules and assembling them into copies of themselves or into other infinitesimal machines that can rebuild an aging heart, a craggy face, a sagging building or fashion, out of a liter of something very much like printing toner, a chair, a table, and the meal that goes on it.

Definitely God territory.

Just ten years ago, nanotechnology was considered only slightly less than delusional—pseudo-science cooked up by physicist Eric Drexler (you'll meet him soon). But it turns out he wasn't delusional. He was right. Building machines the size of molecules *is* possible, as impossible as it sounds. The proof is really all around us. Walk into a redwood forest or your own backyard. Redwoods are molecular machines. They gobble molecules out of the air and ground and gather energy from sunlight. Then, using the molecular machines nature has provided, they transform themselves into massive, majestic, enormously tall trees. Or look at yourself, a molecular machine that walks, breathes, thinks, and feels.

We've generally concluded that this kind of molecular manufacturing is the exclusive province of nature, or God, well beyond human ken. Now, however, some scientists have actually managed to move matter one atom at a time. The next step, move enough atoms precisely enough to build molecular machines, gears, and switches that can then gather other additional atoms to build more molecular machines. (These are called assemblers.) This isn't fundamentally different from what every living organism on the Earth does every moment it is alive. When machines can do this, nearly anything will be possible.

Another area where we are dabbling in godhood is virtual reality. Molecules are small, but digits are less than small, they are ephemeral. Wisps of information that can be combined and transformed into something "virtually" real, but not really real. Or are they?

The ultimate virtual reality machine is the holodeck, *Star Trek*'s very own Garden of Eden. Scientists are making significant progress toward creating something like a holodeck. Entertainment companies must be licking their chops because what could be more entertaining, more lucrative, more "real" than putting yourself in *any* reality you can imagine, with impunity. Be anyone you want, meet anyone you want, shuttle from one era to another—one *world* to another—or hop into places and times that never existed at all, and do it in a blip without ever placing a single mile between you and home. In such worlds we are the creators of the "reality." We can control what happens as surely as the Greeks and Norse knew their pantheon of gods sat omnipotently and omnisciently on Mount Olympus or within the halls of Valhalla and remotely controlled the events of us poor, mortal slobs on Earth.

Which brings us to a final attribute that God has and we don't—immortality. But God knows we badly crave this particular trait, at least if the business that vitamin companies, gyms, spas, and cosmetic firms are doing is any indication.

Strictly speaking, I suppose we don't so much relish immortality as we long for eternal youth. Nobody wants to be 150 years old and look like Whistler's mother. We want to be 150 years old and look like Brad Pitt and Britney Spears or Cary Grant and Rita Hayworth. Immortality is a theme we certainly explored thoroughly in *Star Trek* from Mudd's women to Rayna Kapec. We even brought Spock back from the dead in *Star Trek III: The Search for Spock*.

But those were movies and television. Smoke and mirrors. These days we're talking reality. Long life is on the agenda. Scientists are exploring workable ways to seriously extend our mortal existence and not one of them involves a plastic surgeon named Armando in Beverly Hills. Discoveries that emerge out of the human genome project will certainly make it possible to eliminate scores of genetically triggered diseases. This in itself will extend life considerably. But the mother of all plagues is aging itself; an affliction almost certainly built into our genetic makeup. All living creatures, even Dick Clark, age and die, but why? What little clocks inside of us wind down and then toll the death knell? Why don't we just keep rebuilding ourselves cell by cell as we did as children right on into eternity? When we know the answer to those questions, we might also be able to develop the antidote.

In the meantime, genetic advances will likely make it possible for each of us to clone the key organs that come in so handy when we get up every morning—hearts, livers, lungs, kidneys. Should one start to fail, new genetic breakthroughs will ensure that you'll have a spare available and you won't have to worry about your body rejecting the transplant. After all, it's yours.

Nanotechnology will also play a role in helping to keep us around longer than the standard actuarial table currently predicts. If we haven't nailed the genetic culprit that causes some people to have high cholesterol, for example, your family physician might inject you with a few million nanomachines designed to enter your bloodstream and scour out all of the plaque that's built up in your arteries. After a couple of days, you can start eating fettuccini again!

All of this has helped me see that there is one point that evolutionists and creationists can agree on after all. Evolution has no plan. God does. Gods, as a general rule, are proactive. "Let there be light" is not a passive statement. God may move in mysterious ways, but he also moves in deliberate ways. And that's the point of the chapters that lie ahead. They explore very *Star Trek*-ian technologies that may soon put us in the position to *deliberately* and *consciously* intervene in the natural course of evolutionary events.

Now maybe this emerging ability is just another twist in the evolutionary road, or maybe God wants company, but there is little doubt that we will soon be faced with some of the same powers that divinities possess. This (gulp!) also means that we will soon face handling the same responsibilities as well. Are we prepared?

I think we may find it complicated business acquiring the power of the gods. The technologies we develop, being double-edged, can cut for ill as well as good. Fire can burn as well as illuminate the darkness.

Gods, the world's primary religions teach us, are wise and

patient, loving and understanding, as a group. Considering the kinds of powers we humans may soon be assuming, I can see how these attributes might come in handy.

So, being among the technologically challenged, let me sound a note of warning: we might want to be careful about what we bring into the world. It's not just a question of playing God simply because we have the power and know-how to—as Captain Picard was so fond of saying, "make it so"—it is also a question of making sure we can handle the job well in the first place.

Nanotechnology offers . . . possibilities for health, wealth, and capabilities beyond most past imaginings.

—K. Eric Drexler

It's the little things that mean a lot.

—Anonymous

18

NANOTECHNOLOGY-OLOGY

It was the very end of 1959 at the California Institute of Technology when the legendary Dr. Richard Feynman gave a fascinating little talk at the annual meeting of the American Physical Society. Feynman, whose name comes up a lot when discussing cutting edge science, is generally considered one of the great physicists of the twentieth century; a Nobel laureate who not only worked on the Manhattan Project, but who, forty years later, figured out why the *Challenger* space shuttle blew up. Feynman had a life-long habit of seeing things from odd and revealing angles and he didn't disappoint the roomful of colleagues he was addressing that night. His talk was entitled "There's Plenty of Room at the Bottom."

During this talk Feynman drilled—with impressive prescience and the puckish humor he was well-known for—right to the heart of the question he had raised.

"Why can't we manufacture small computers somewhat like we manufacture the big ones?" he asked. "Why can't we drill holes, cut things, solder things, stamp things out, mold different shapes all at an infinitesimal level? What are the limitations as

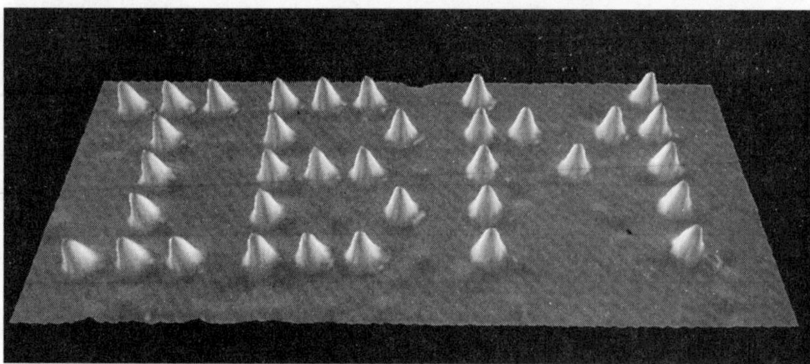

Using a scanning tunneling microscope (STM), scientists at IBM spelled the name of their employer one xenon atom at a time in 1990. This illustrated that atoms can be manipulated at the nano level. As Richard Feynman had predicted in 1959, the laws of physics would not prevent it. The questions now: Can we learn to assemble machines at the molecular level, and can they go on to make other molecular machines?

to how small a thing has to be before you can no longer mold it? How many times when you are working on something frustratingly tiny like your wife's wristwatch, have you said to yourself, 'If I could only train an ant to do this!' What I would like to suggest is the possibility of training an ant to train a mite to do this. What are the possibilities of small but movable machines?"

Before he was done talking he let everyone know that the possibilities were pretty good. "The principles of physics, as far as I can see," he said, "do not speak against the possibility of maneuvering things atom by atom. It is not an attempt to violate any laws; it is something, in principle, that can be done; but in practice, it has not been done because we are too big."

With this speech, Feynman foresaw the possibilities of what we today call nanotechnology, or more precisely, molecular engineering.

Star Trek might have anticipated cell phones and personal digital assistants, voice recognition and CAT scans, but when it came to nanotechnology, we never saw it coming. Well, it is true that the androids invented something called a nanopulse laser in

"I, Mudd," but I think the writers were just trying to come up with a cool name for a piece of technology.

The irony is that the world that *Star Trek* envisions can never come to pass without the development of something at least *like* nanotechnology. Without it there could be no Data, no holodeck, no warp navigation, deflector shields, or food replicator, and certainly no beamings-up. The monstrous computing power needed to run the *Enterprise*, even the alloys and structures that hold the ship together, would almost certainly require nanotechnology. You simply can't accomplish all of this kind of apparent magic unless you are manufacturing and computing at the very fine level of one . . . atom . . . at . . . a . . . time.

But there's a good reason nanotechnology never came up in the original series. The idea as it is now conceived simply didn't exist at the time! In fact, the series predated a fully formed version of the concept by some twenty years. It's true, Richard Feynman had asked questions about molecular engineering and he tossed the idea out there, very generally, in 1959, but it was a young engineer named Eric Drexler who first pulled together a detailed scientific vision of how nanotechnology might actually work. In fact it was Drexler who came up with the term. But he didn't come up with it when I was doing battle with the Horta and jumping through time tunnels in the mid-sixties because he was in grade school then. In fact by the time *The Next Generation* went on the air, Eric had only recently published his first papers on the subject. Today the science is so well accepted that the federal government allocated nearly one to three billion dollars for nanoresearch through 2002.

Dwarf Science

Nano, I am told, comes from the Greek word meaning "dwarf," which is fitting since anything that takes place at nano-scale is exceedingly small—the atomic level to be exact, where the dimensions of objects are measured in billionths of a meter, or a nanometer. How small is a nanometer? Go out in your backyard and look for the tiniest bug on the tiniest blade of grass. Now fly up that bug's nose on the back of a microscopically small dust mite.

You are still ten million times larger than a nanometer!

As of this writing, engineers can place one hundred million transistors on a sliver of silicon that measures one-tenth the surface of your fingernail. One hundred million! And in another year, it'll be two hundred million. Small, right? Yes, but gargantuan when compared with nano-scale dimensions. Say you took this one hundred million transistor slice of silicon and reduced it a hundred times, you'd think that would get you to nano-scale, right? Wrong. True, you would start to see molecules themselves at this level, but they would still be very distant, like looking at basketballs from a hundred yards away. If you reduce the chip one thousand times, you finally, at long last, get to nanoland, a place where DNA molecules are busy writing the code that makes all of life on Earth possible.

Or look at it this way. If we scaled one of the transistors that have been crammed onto that chip up to the size of this page, a single atom would be slightly smaller than this period right here →.

More than one scientist has told me that in the future we will look back at the tiny silicon wafers we use today and wonder how we managed to get anything done with such big, elephantine chunks of sand. Even a human cell is ten thousand times the size of a nanometer! This difference alone is like comparing Godzilla with Thumbelina.

We are talking small.

Really small.

Little Things Mean a Lot

Small, however, should not be confused with insignificant. Everything that happens in the universe happens first in nanoworld because all matter—solid, liquid, or gas—is constructed of atoms. All of the elements in the periodic table (hydrogen, iron, zinc, oxygen, etc.) form nature's LEGO set. Linked together in yet other ways they give us everything from drops of water to lumbering elephants. Whether it's the accretion of atoms that makes planets possible, or the operation of our immune systems, it all starts in nanoworld. Of course we only figured this out recently because we can't *see* what is happening at this atomic scale. It's *way* invisible. We can only see the accumulated results.

And the results can be pretty impressive. Stars, like the sun, are one. Atomic fusion takes place at nano-scale* and produces the power that has made life possible on Earth. Blue whales are another—enormous structures, a hundred feet from stem to stern, that swim the oceans. They have been built (and are constantly being rebuilt) by legions of natural nanomachines called ribosomes that are controlled by their DNA. So are pigeons and lilacs and all of the corn in Iowa. *You* are the result of nanotechnology. You are a rippling mass of five trillion (that's 5,000,000,000,000) cells. Various cell-groups in your body patiently break down the supersized fries you just ate into their molecular constituencies. They then send them off to other cells that use chemical reactions and enzymes to reassemble them and turn them into bone and blood and neurotransmitters so you can read this passage and understand it. You *do* understand it, don't you?

---

*Strictly speaking, fusion takes place at the scale of the nuclei of atoms, so even though it's called atomic power, it's actually not taking place at nanoscale, it's taking place at an even smaller, subatomic scale!

This is impressive work when you consider it is all done at very high speed, very low energy, and a few atoms at a time. It's hard to believe you could build a giraffe this way. But that's the way they're built.

Eric Drexler has pointed out that there are some crucial differences between his version of nanotechnology and nature's. Natural systems can't be programmed to do what we want them to do, at least not very easily. (Although over at the Human Genome Project and at several large drug companies, they are working on this.) On the other hand, the kinds of nanomachines that Eric has dreamed up *can* be programmed. In fact, that's their greatest strength—theoretically they can be told what to do, very precisely.

And if this happens, some very *Star Trek*-ian things will follow.

The Nano Man

Naturally, Chip and I realized we needed to see Eric Drexler, the Nano Man, the undisputed father of molecular nanotechnology himself. We're on a mission, right? So we headed off to the Foresight Institute in Silicon Valley to meet Eric face-to-face. Christine Peterson had gladly arranged a get-together with Eric, Ralph Merkle, and herself. Chris is the Institute's president, Eric is chairman, and Ralph is one of the world's leading nanotechnology experts.

The name Foresight Institute might conjure up images of a gleaming glass-and-steel skyscraper with a floor of offices brimming with the latest technology and the most fashionable dot.com furniture, but that would be wrong. Instead Foresight resides on a quiet side street in Palo Alto. It consists of a small group of ivy-covered, one-story offices that would probably be more at home in Bangor, Maine than the belly of the Great

Silicon Beast. The offices are plain, the furniture used, and the computer equipment functional. The thinking, however, is glistening, shiny and beyond modern.

We arrived one clear, brisk California afternoon, and sat down with everyone in a conference room outfitted with a nice big white board and lots of markers so that when I became confused everyone could draw nice, simple pictures for me.

Eric Drexler can look bookish, especially with his glasses on. But he couldn't care less. There is nothing of the manicured, Silicon Valley-ite about him. He's the first to admit that he's a devoted academic who would just as soon attack mind-bending math problems as eat. Nothing better than a good, juicy quadratic equation to sink your teeth into (preferably rare). This lascivious desire to mathematically strip engineering mysteries of their secrets is what first set Drexler on a collision course with the outrageously small.

Back in the 1970s he received his undergraduate degree at MIT in interdisciplinary science (already showing a tendency for maverick behavior). He quickly followed with a master's degree in engineering and then, finally, a doctorate in molecular nanotechnology, the first ever. (Guess who was his supervisor? Marvin Minsky. The future seems destined to be littered with the students of Marvin Minsky.)

Drexler also had an interest in space colonies and satellites and creative ways of transforming the raw material in asteroids into human-made, precisely engineered machines like spaceships and satellites. For a while he worked with Gerard K. O'Neill, a tremendously creative scientist at Princeton University who invented and developed the storage ring technology that became the basis for all high-energy particle accelerators. Later, through a nonprofit organization called the Space Studies Institute, O'Neill developed truly visionary plans for the settlement of space.

He didn't mess around with piddling little NASA-style space stations either, according to Eric. He foresaw mass migrations into space made possible by fleets of orbiting, rotating miniworlds complete with sunlight, gravity, farmlands, and small cities. These places wouldn't be measured in yards but miles and he foresaw millions of Earthlings making their homes in colonies like these. If you were walking down a path within one, it would be so large that it would feel just as though you were walking down a country path in Kentucky or Pennsylvania or among the rolling hills of southern England.

Of course when constructing space settlements of this kind, you need some serious building supplies. You can't launch the whole kit and kaboodle from Earth like we are currently doing with the International Space Station, as if it were an oversized Tinker Toy. If you want to build anything on a large scale, observed O'Neill, you needed the stuff that dreams like this could be made of.

Asteroids.

Thousands of asteroids roam the vicinity of Earth, it turns out, each loaded with all of the raw material you could possibly need to build Earth-like outposts. At least that's the way O'Neill saw it.

The problem with asteroids is that the natural forces of the universe just haven't put all of the raw material, all of the nickel and iron, hydrogen and oxygen, into quite the right configuration for living, says Eric. Earth is a freak of nature, at least in this parsec of the galaxy. All of the right elements have fallen together in more or less the right way here. But asteroids—that's a different story. So, O'Neill wondered, how could you reconfigure the little lumps of floating rock into tiny Earths?

Drexler says he began to think it was a matter of scale. "Asteroids, like everything else, are just big blocks of randomly assembled atoms and molecules. If you could find a way to rearrange those atoms, you could turn the asteroid pretty much into anything you wanted; even something livable. You would just have to do some extremely fine scale manufacturing."

However, to do that, you would have to create very small machines, infinitesimally tiny machines, dwarfish machines, in fact. Maybe even machines the size of molecules. And as anyone in his right mind knew, *that* was insane.

There Really Is Plenty of Room at the Bottom

Except for Eric Drexler . . . and that was partly because he was familiar with the outrageous possibilities that Richard Feynman had discussed during that little known talk in 1959. Drexler knew in the early 1980s what all scientists know now and what Feynman suspected forty-three years ago: that there isn't *any* reason you can't move matter one atom at a time. No laws of physics countermand it. A few years ago, in fact, using an electron-tunneling microscope (essentially a stick with a point on it no wider than an atom), researchers at IBM positioned atoms beside one another so that they spelled (surprise!) "IBM." Since then several others have done the same (except they have usually spelled something other than IBM). Impressive as this is, it's still a long way from moving the little buggers precisely enough that we can build something as useful as gears and switches. We're in the Stone Age stage of nanotechnology, admits Drexler. But that's changing fast.

Inspired by Feynman and O'Neill and his own long fascination with less-than-obvious-science, Drexler says he sat down and wrote a book called *Engines of Creation*, published in 1986. "The book outlined what the world might really be like if nanotechnology was truly possible," he says.

In it he explained how the human race has pretty much been rearranging molecules for as long as it has been making tools.

Flint axes, for example, were shaped by whacking two rocks together and then chipping one away with the other in increasingly fine motions to create a sharp edge.

"Our ancestors grasped sharp stones containing trillions of trillions of atoms," he points out, "and removed billions of trillions of atoms to make their ax heads. Later we learned to move other atoms and molecules around by cooking rock and melting ore out of them. Copper, for example. Once the copper cooled, Bronze Age smithies would then pound and hammer the trillions of copper atoms into swords or plowshares or a nice goblet."

Unknown to the smithy, says Eric, he could do this with copper because its atoms stick together in regular patterns (when cool enough) and slide over one another while still remaining connected. Glass molecules, on the other hand, don't do this so well. Hit them with a hammer and they fly apart rather than slide. Rubber, it turns out, acts the way it does because it consists of strings of molecules that kink up, and entangle one another like a series of interconnected springs. Pull rubber in one direction and the springs stretch. Let them go and they recoil into their original position.

The point is, explains Dr. Drexler, things act the way they act because of the way those trillions of invisible atoms and molecules are structured. Assemble carbon atoms one way and you get diamonds, assemble them another and you get coal. Put sand molecules together this way and you get a sandcastle, another and you get a computer chip that can do ten million operations per second.

Who was it that said the devil's in the details?

Now, Drexler asked, just step back, way back, and think for a minute. If you could assemble any number of molecules any way you wanted, there would be very little that you couldn't do. You could essentially perform something very close to what the ancients called alchemy—the transmutation of almost anything into anything else. Not only that, but the transformation would take place on an extremely fine scale, so that whatever you created could be stronger, faster, lighter, you name it. It would be magical! There would be applications in medicine, communications, manufacturing, space exploration, computing.

Life would be good. That's the promise of nanontechnology.

Building the Future . . . One Atom at a Time

The world that Drexler laid out in *Engines* is remarkably *Trek*-like in this sense: It's optimistic without being utopian or Pollyanna-ish. Economies would be in for tectonic shifts as the cost of manufacturing dropped through the floor. Energy needs would change dramatically because efficiencies would increase enormously. Look at how little energy it takes to run something as complex as you or me. Put in some meat, fish, fruit and vegetables, maybe an occasional apple pie à la mode, and you get walking, talking passion and creativity. Pollution, at least on the scale we see now, would disappear. And should an *Exxon Valdez*–style oil slick once again be inflicted upon the environment, we could create battalions of nanomachines specifically built and programmed to eat oil, rearrange the molecules, and transmogrify it all into a harmless by-product.

Or we could create outrageously powerful computers—computers that easily would have the horsepower to run warp drives, calculate trajectories at the speed of light, and handle the complex machinations of creatures like Data. Using nanotechnology, we could build Data's positronic brain (if only I knew what a positronic brain was!). Enormous ships, colonies and Deep Space 9s would only be a matter of time. In nanoworld, Deep Space 9 might, in fact, have started its career as an asteroid, transformed atom by atom into the hotbed of interspecies activity that made the show so popular.

There are, of course, applications in medicine that McCoy would have relished. Remember all of those I'm-a-doctor-not-a . . . remarks that he used to make? He liked to point out how primitive the medicine of the twentieth century was when doctors actually cut patients open like slabs of meat and sewed them up as if they were so much sailcloth. Traditional as he was, Bones would have embraced nanotechnology with grinning enthusiasm. Nanotechnology, according to Eric, would have allowed him to use a molecular scanner that was so powerful that it could peer into any body and detect whatever ailment or injury he needed to track down.

It would also make it a relatively simple matter to repair the damage. All he'd have to do is hypospray a few million nanomachines into the patient's bloodstream and have them get to work repairing the ailment molecule by molecule, something like a supercharged, made-to-order immune system.

Despite all of these wonderfully optimistic, Roddenberry-esque scenarios, however, a nanotechnologically enabled future would not be without its dangers. Drexler is the first to admit it. This is powerful stuff and he says we will have to be very careful. There's the "gray goo" problem, for example.

The Gray Goo Problem

Imagine being able to build a machine that can build other machines just like it; a machine that can assemble an unlimited supply of its own identical twins right out of the material that's sitting around it. Now imagine that this machine is extremely small. What would you call this? On the one hand you might call it the DNA (deoxyribonucleic acid) in each of your cells. DNA is what makes it possible for all of our cells to replicate.

This keeps us from turning into little piles of dust. On the other hand, you might also be describing a deceptively powerful, extremely dangerous, and entirely invisible weapon.

Just for fun let's assume that it's a nanomachine you're describing, and you happen to have such an infinitesimally small gadget hanging around (slightly more than one hundred billionths of a meter from top to bottom). And let's further imagine that it is programmed with very simple instructions: Go herd up a bunch of other atoms and molecules and assemble them in exactly the same way you are assembled. If each newly minted machine did this, and then each subsequent generation of machines also did this, then the Tribble Effect would kick in and before you know it, you'd quickly have a whole planetful of nanomachines making yet *more* versions of themselves. This is called self-replication and it can accomplish real miracles—you and I and the rest of life on Earth being the most impressive current examples.

However, this sort of thing can also get out of hand. The exponential growth of the world's population (now at six billion and climbing fast) is the dark side of the miracle of self-replication. Then there's tribbles, and we all know what happened there. Now imagine that in the nanofuture a terrorist creates a rampant, self-replicating nanomachine that begins constructing bazillions of copies of itself. And let's say that this particular machine's job is to disassemble, say, carbon molecules so it can make more versions of itself. Well, almost *all* living things are constructed of carbon molecules, from trees to elephants to Jay Leno. So what these machines would merrily begin to do is go about disassembling everything all around it like one of those flesh-eating strep infections that *The National Enquirer* likes to feature so prominently on its covers.

This would be a bad thing.

In *Unbounding the Future*, another book that Drexler

wrote—along with Chris Peterson and colleague Gayle Pergamit—they call this, of all things, the *Star Trek* scenario. Why? Because the out-of-control self-replicating machines refer to a *Next Generation* episode entitled "Evolution," in which a crowd of nanites—nano-robots in *Star Trek*–speak—gets loose and nearly wipes out the *Enterprise*-D lock, stock, and plasma manifolds.

Turns out that not only were these nanomachines smart, they were also sentient! Nothing less than a race of molecule-sized, homemade aliens. Lucky for them. Since they were conscious and intelligent though very small, they were spared eradication. Not only that, but they were instead granted colonization rights on planet Kavis Alpha IV where presumably they turned their new home into a nanite resort and made billionths of dollars.

The reason the nanotechnology community calls this kind of behavior the "gray goo" scenario is because anything these little machines come into contact with would quickly be transmogrified into gray, utterly lifeless, goo.*

"The gray goo scenario might unfold like this," says Eric. "Down the road, someone could engineer a machine capable of goo-ifying everything; a superorganism that is bacteria-sized, omnivorous, and able to survive in all sorts of extreme environments. You wouldn't think that anyone would do this *on purpose*, unless they were insane, so presumably we'd build in some safeguards; a command, for example, that says, 'Stop replicating after you've made two copies.' Or, 'Don't make copies if you are outside of this solution.' "

But as we know, mistakes can happen when copies of molecular-sized machines are made. In fact, these mistakes are what drive evolution. If there had been no mutations in DNA, no

---

*Actually, Eric says, whatever was left would probably not be either gray or goo, but it wouldn't be very useful, so you get the idea.

unexpected changes, we wouldn't have evolved and the planet would have about as many life-forms as politicians have good ideas (maybe two). But evolutionary change moves very slowly, and the world had time to adapt to its innovations. The sloth with which evolution advances makes everything around it capable of adapting.

But if the safeguards in a few highly destructive nanomachines broke down, well, in the immortal words of a character in *Aliens*, "Game over, man!" "We'd be faced with an ecological disaster," says Eric, "that would make the largest oil spills, the most damaging forest fires, or the most titanic tsunamis look like extremely small potatoes."

Of course the upside of this, Eric points out, is that making destructive nanomachines like these isn't easy. More to the point, right now they are impossible, but even in the future, when we have the technology in hand, it won't be easy. It would take a lot of effort. You would basically have to want to do it on purpose, and that makes no more sense than building a nuclear bomb and then taking it out to your backyard and detonating it.

"We understand that all technologies are double-edged," says Eric. "And we understand that when dealing with fire this hot, it's probably best to be thinking now—decades before we are really capable of making deadly stupid mistakes—how we plan to avoid complete, planetary meltdown.

"Understanding that these technologies can be used for destruction and oppression and other sorts of abusive applications is important. On the one hand we can do a tremendous amount of good; on the other we can't avoid the negative applications by stopping the technology because there's no one in a position to do that. It's just not an available option. History teaches us that if a technology *can* develop, it *will* develop. And so we're pretty well stuck with moving forward and trying to avoid the bad while pursuing the good."

The truth is we could dream up endless nano-scenarios, both wonderful and horrifying, but it still doesn't address the basic question: How in the world would you ever create nanomachines, preferably the good kind? What does it really take to put the tech in nanotechnology?

Going Nano a Nano

"Okay," says Eric, running his slender fingers through his hair, "here is something that could be built that people might arguably want in a world where molecular manufacturing is possible.

"Imagine a box about the size of a microwave oven. It has little rubber feet on it and a fan and it plugs into the wall, like a copying machine. And it has a button on it that says START. It's right next to the tag that reads ACME HOME MOLECULAR ASSEMBLY MACHINE. On the little digital display panel are the words PRINT CATALOG.

"So you put some stuff in here [he gestures like he's pouring oil into a car engine]—the equivalent of nano-toner—and you hit the START button. [Now he makes a whirring mechanical noise.] It generates some heat. It's not perfectly efficient. But after a while [another noise, this time like a *ding!*], you pull out a catalog. But it's not just a paper catalog; it's a flat screen device that has a high resolution, full color, 3-D screen. This thing has more storage capacity than the Library of Congress because it's a molecularly engineered display gadget.

"Anyhow, you can sort of leaf through it, or maybe surf through it like a Web page, and select some item you like. Maybe it's a GPS unit that flips out of your watch and has a really nice map and if you want you could unclip the map and hold it up like a monocle and see it overlaid over any part of the world you wanted—whatever. All of this is described in the catalog. Maybe it even talks to you, describing the entire gadget. So you say to yourself as you stand in your kitchen, 'Yes, I'd like one of those.' And the unit says back, 'Well, Bill, that doesn't have very much mass so I can make one up for you without you having to pour any more toner into me quite yet.' And you say, 'Okay.'

"Now the machine whirs for a little while longer and then *ding!* it's done. Just like that! You open the door and there's your wristwatch-sized GPS system ready to go. What did it cost you? Well, I don't know what the software cost you. That depends a lot on intellectual property. But the energy needed to make it was minuscule. It came right out of the wall socket and the gadget was small, just several trillions of atoms, so it didn't take much power. And raw material? The toner was cheap, probably cheaper than fax toner is today because fax toner is kind of fancy stuff. I mean this thing will use nail polish remover kinds of chemicals, and turn them into computers and GPS systems."

I almost asked if this gizmo would work better than the one in my rental car, but before I could Eric was already elucidating more of the future.

"And you can also make some other things with it. Maybe you say to yourself, 'Oh, I would like a roll of that tough-sticky-on-one-side-high-efficiency-rather-dark-solar-cell material.' And so you pour in some more toner and *ding!* you pull out this roll of stuff that's kind of like trash bag material. A big roll. And let's say you spread it out over the surface of your driveway. Now this stuff is tough and you can drive over it and it's kind of tasteful (black is always tasteful) and it absorbs light and converts fifteen percent of it to electricity. And there's a little wire and plug at the end of it. You can plug it into your power supply. Now it's not *using* power, it's *generating* it."

"Now if you wanted more raw material for your ACME Home Molecular Assembly Machine, you could take carbon right out of the atmosphere to make more 'toner.' I mean there's plenty of excess CO_2 everywhere. You can filter that out of the air, clean it up a little bit, and turn it into all kinds of useful objects."

"You mean make things literally out of thin air?" I ask, scratching my head.

"Absolutely," says Eric, as though what he has just described is the most normal thing in the world, which it is, I suppose, if you live in *his* world.

"Now if you are personally concerned about people who are living in an impoverished village in Bangladesh, and you feel it would be nice to spread the wealth, then you could select another one of these Molecular Assembly boxes. And have it generate another identical ACME replicator all folded up. And when you unfold it you would see right on the display panel PRESS BUTTON FOR CATALOG. But, of course, you'd probably want to fold it back up, stick a stamp on it, and mail it to Bangladesh."

He smiles a smile reminiscent of Mona Lisa.

I just look at him across the conference table like one of those cartoon characters whose jaw has hit the floor. Then, once I've put my mandibles back in place, I say something really pithy like, "Yeah, all right, but what's going on *inside* the box?"

At this point Ralph Merkle speaks up. Ralph is about the most affable, outgoing scientist on Earth (big *Star Trek* fan too). In addition to his brains, he brings this nice baritone and a barreling, infectious laugh to whatever he does. Here he was, just having rolled in from Vienna where he had been giving talks on nanotechnology, and he's as energetic and fresh as a thirteen-year-old on his first date. Like Eric, he has a way with words and a nice knack for explaining complex science in ways that even I can understand.

"Well, a whole lot of little tiny machines working real fast with a lot of software are assembling whatever you asked for, that's what. They are turning the toner into whatever you ordered, one atom at a time."

How?

The machine, he explains, organizes atoms from the nano-toner you poured into tiny molecular assemblers, weenie machines a hundred billionths of a meter from one end to the other that go on to assemble other things that are bigger than they are.

To do this, he says, the machine needs a detailed description of what you asked it to make, a program of sorts, which the machine then uses to create not information like your average computer, but actual real, three-dimensional objects. Just like *Star Trek*'s replicator. It places these atoms here, next to these and assembles molecules, and then these molecules go here next to these until an actual, visible . . . thing . . . begins to form. You know, the knee-bone's connected to the leg bone. That sort of thing.

"In other words," I say, "you build an invisible, computerized robot that is programmed to build something, molecule by molecule."

"Yes," says Eric. "And this is possible because a nanocomputer, once such a thing exists, could hold the guts of a trillion of today's desktop computers."

"A *trillion* desktop computers," I mumble, dumbfounded. "A *trillion?* Tell me more."

Ralph leans back in his chair as if this were the easiest thing in the world to explain. "Today's manufacturing methods are pretty crude at the molecular level," he says. "We cast, we grind, we mill. Even the lithography we currently use to create minuscule computers chips, moves atoms in great thundering herds. It's like trying to make things out of LEGO blocks with boxing gloves on your hands. Sure, you can push the LEGO blocks into big heaps and pile them up, but you can't really snap them together the way you'd like. Nanotechnology lets you snap everything together *exactly* as you want."

"And there is snapping going on inside the ACME countertop?"

"Precisely. At least theoretically," he says, walking up to the white board. "You start with the small things. In fact, the smallest things you can start with—atoms and molecules. They are then fashioned into little three-dimensional parts about the size of a nanometer. [He draws a box.] And these parts are picked up by little robotic arms. [He draws little arms. No offense to Ralph, but all of you artists out there, your jobs are secure.] We call these gadgets 'assemblers.' The little robotic arms now take the one-nanometer parts and they use them to build other parts that are *two* nanometers in size. Then you take the parts that are two nanometers in size and use them to make still more parts that are *four* nanometers in size and so on and so on . . . until, pretty soon, you get, well, whatever it is you wanted to build in the first place." (See Ralph's Web site for more information on how assemblers work: www.merkle.com.)

Okay, now I'm getting it. This is obviously what Feynman meant when he said he wanted to train an ant to train a mite to make very small but movable parts, except here you have mites building ants so the ants can build something bigger and so on.

And this sort of magic, this alchemy, is how far off? I ask.

"Not impossibly far," says Eric. "People sort of already know what to do. It's like space exploration. It took years and years and years and years and years before anybody managed to get anything into low Earth orbit. Developing the right rocketry was the trick. Then the next thing you knew, we had space probes zipping by Mars and Venus. Then we started sending things out of the solar system.

"The point is there was a threshold to cross, and then once

we crossed it, it was a modest reach beyond low Earth orbit to actually taking things to a whole new level. So we knew the principles of space flight a hundred years ago, but it took a while to figure out the details of *how* to actually transform the knowledge into working technology, and then we made it happen."

But how long before we can do *this?* 2010? 2020? 2050?

"2011," says Drexler, grinning.

"I'm very evasive on that question," says Ralph. "Too many variables and not all of them have to do with the science."

Well, that seems true. There are cultural, moral, legal, financial issues—the stuff of human affairs. After all, if it hadn't been for the Cold War, we might never have launched anything more than a few weather satellites into space. Nevertheless, I persisted.

"Why don't you just say 2007?"

"Well, maybe that's not a bad approach," booms Merkle. "If you say 2013, July 17 at 8:17 in the morning, that's just as good as wiffle-waffling around."

"Great!" I say. "Damn, I can't wait until 8:17 A.M.! In fact, I'm going to be there at 7:00!"

"It could stabilize things," says Drexler, grinning again.

"Okay then," I say, hitting the table. "It's agreed. July 17, 2013. That's the date, and you heard it here first."*

So mark your calendar. Check CNN from time to time; see if the complexities of human affairs gets in the way. And keep your

---

*If you're really interested in making certain the human race keeps its date with nano-destiny, you can enter the competition for The Feynman Prize. At the end of his talk back in 1959, Feynman himself offered a prize of $1000 to anyone who could figure out how to build a working nanomachine; just for the fun of it, he said. $1000 was a considerable sum in those days, but does not carry much water forty years later, so the ante has been upped. The Foresight Institute is now offering $250,000 to the creator of nanomachine #1. For more information, log onto the Institute's Web site at www.foresight.org/GrandPrize.0.html.

eyes on the technology, the nanotechnology, because it is moving along apace. In fact it's happening right now, even as you read the words before your very eyes.

Nanotubing

For example, take carbon nanotubes. It's true, strictly speaking, they aren't the self-assembling nanomachines that Drexler and Merkle talk about, but to scientific rubes like us, it's close enough, and they do represent strange and amazing things taking place at the molecular scale. Look at them as a first stage assault on a full-blown nano-age. That's the way researchers working on them see them. The interesting thing about carbon nanotubes is that they have chameleon-like, almost magical, properties that promise to make them very damned useful pretty damned soon.

But what, you are wondering, is a nanotube, besides a tube a billionth of a meter long? Here's a meager explanation based on information that far more knowledgeable people than I have imparted to me. You make a nanotube by linking carbon atoms up into hexagons. Then you hook all of those together into something that looks like a rolled up bale of chicken wire. (The tubes will self-assemble in this form under certain extreme conditions so you don't need any Drexler-machines to assemble them. This is just a fluke of nature.)

What good is a bunch of miniature chicken wire? Don't be a smart aleck! Researchers have found that if you line up a row of carbon molecules on a tube in a straight line, they act like metal and will conduct electricity; this even though they are not metal. That's one chameleon-like attribute.

Now let's say you twist the hexagons into a helix (the shape, coincidentally, of DNA). When you do, they start to act like semiconductors, the stuff that computers are made of. (Semiconductors insulate at low temperatures and conduct at high temperatures which enables them to go both ways and complete circuits that represent on and off, or zero and one.) Connect these two shapes in different ways and you get some strings of molecules that act like transistors and others that act like diodes. In other words, depending on the way you shape the carbon atoms, you can have all of the "guts," as Eric would say, of a very, very, very small, yet unbelievably powerful computer.

In addition to these quick-change attributes, nanotubes are about six times lighter and ten times stronger than steel. This might not be quite the wonder metal that *Star Trek*'s duranium is, but, hey, we're only in the twenty-first century so give us a break. Add to this the ability of nanotubes to also act as extremely small, stupendously powerful computers and you can see we will soon be able to build outrageously strong materials that are also highly intelligent. Computing power, lots of it, would be built right into the objects themselves—cars, buildings, roads, jets, nearly anything you can imagine. Your ride to work wouldn't simply be a box filled with computers, it would be *made* of computers; yet it would also be extremely resistant to being flattened (with you inside of it) by a sixteen-ton trailer.

You can begin to see how all of these converging technologies could then lead to everything from highly advanced space vehicles to the outrageously clever robots that Hans Moravec foresees. I mean how often do you get something that is not only unbelievably strong, but seriously intelligent as well?

Maybe in certain members of the NFL.

Whatever the case, just for starters, just to sort of limber up the possibilities, researchers today are looking at using nanotubes to create very cheap, low-powered, flat-screen computer displays (full color); or amplifiers for communicators . . . er . . . cell phones, or ion storage for batteries that could power electric cars. Every one of these represent multimillion-dollar markets. And this is just the first wave of innovation!

The big push, the big nano-application that everyone is eyeing, is computing. That's what scientists at NASA and MIT tell me. Why? How about lots of money?

First, the demand for smaller and more powerful computers is unlimited and will be until God calls a halt to human curiosity.

Second, the current method for cramming transistors onto silicon chips will hit the wall around 2015. After that it will no longer be possible to etch transistors on silicon using light.

Third, in the same period of time, the cost of building microprocessor fabrication plants will have reached nearly $200 billion . . . per plant!! That's a lot of bucks, even for Intel. Something new is going to have to come along or else the cost of chip making will outpace the income of some pretty large nations, maybe even Bill Gates.

We can't have that.

Whatever the case, if we actually manage to attain the magical goals that nanotechnology in all of its forms could make possible, if we can truly begin to train "an ant to train a mite" and begin assembling Drexler-machines, then we will very likely see outrageously advanced *Star Trek*-nologies unfold. But they won't unfold in three hundred years, they'll unfold in a generation—ten times faster than Gene Roddenberry thought.

Suddenly the positronic possibilities of Data's brain will become real. The computing power needed to do all of those mind-bending transporter calculations will not be so outrageously difficult. Talking computers and small, nimble robots will be a snap (well, maybe not a snap, but way easier to create than they are now). And the medical miracles practiced by Dr. McCoy? They'll be amazing, but not miraculous. Simply put, much of what *appears* impossible today will no longer *be* impossible.

In fact someday, not very far down the road, battalions of nano-assemblers may begin work in Earth orbit on a ship designed for interstellar flight. A ship assembled atom by atom, made of unbelievably strong alloys, rippling with unthinkable computing power, may prepare for a voyage, bound for parts unknown. Perhaps, just like certain sailing frigates and aircraft carriers and space shuttles that preceded her, we'll name her the *Enterprise*.

I like that name. It has a nice ring to it.

Do not go gentle into that good night,
Old age should burn and rave at close of day;
Rage, rage against the dying of the light.

—Dylan Thomas

It's not that I'm afraid to die. I just don't want to be there when it happens.

—Woody Allen

19

GIVE ME IMMORTALITY OR GIVE ME DEATH

In the beginning . . . of this project, I mean, I had planned to feature a section on *Star Trek* medicine. After all, advances in medicine have been so rapid, and McCoy was such an important part of the original series, it seemed to make sense. You'd have to agree that the nature of his magical ministrations was one of the real appeals of *Star Trek*, especially in those early days. I can't pinpoint why, exactly. Maybe it was just that weird warbling sound his little faux instruments made as he waved them over your body, or maybe it was those cockeyed eyebrows that DeForest used like facial exclamation points. Or maybe it was simply the idea that you could perform medicine that looked suspiciously like magic.

This last "maybe" is, of course, another sign of *Star Trek*'s prescience because this is exactly the direction in which medicine is headed. The days of the scalpel are numbered. Already we're using all sorts of noninvasive technologies to explore injuries and outfox disease. CAT scans, magnetic resonance imaging (MRIs), ultrasound, genetic engineering, chemotherapy, radiation treatment, arthroscopic surgery. And there's plenty more around the bend.

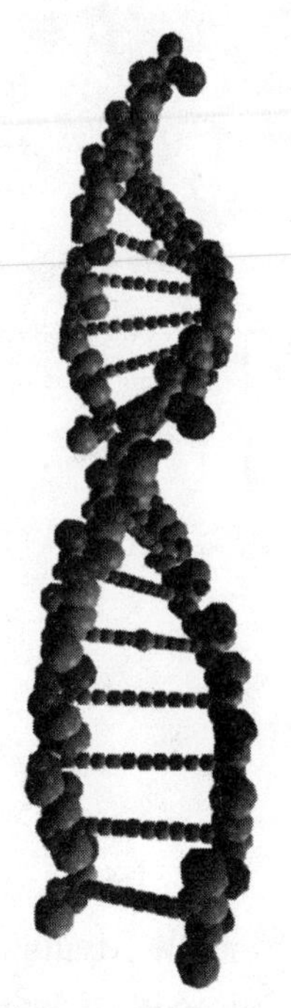

The long, linked ladder of DNA that makes each one of us, and all of life on Earth, possible. We are learning the recipe, but it hasn't been easy. *(Doug Drexler)*

Back in neolithic times, when the first *Star Trek* episodes were being shot, there were no CAT scanning machines, or MRIs. It would be more than a year before Dr. Christaan Barnard performed the world's first human heart transplant. Cloning was pure fantasy, in fact it had only been a touch longer than a decade earlier that James Watson and Francis Crick had even revealed the secret structure of deoxyribonucleic acid or DNA, that string of molecules responsible for life on earth.

In the time that has passed since then, we humans have increased the life expectancy of the average American from sixty-eight to seventy-eight years of age, and it's still climbing. A recent study reported that predictions of life expectancy by 2050 were low. Now it seems that the average American born in 2050 will live eighty-four years. My guess is that as we march into the twenty-first century this number also will turn out to be low. Way low.

Why? Because as difficult as this is to comprehend, when you look at the breakthroughs we've been investigating just in this book (which is far from exhaustive), you can see we will soon evolve an entirely new kind of medicine; a McCoy kind of medicine that will make the advances we saw in the twentieth century look as barbaric as bloodletting and the treatment of bad humours in medieval Europe.

And this is why, despite my initial inclinations to investigate

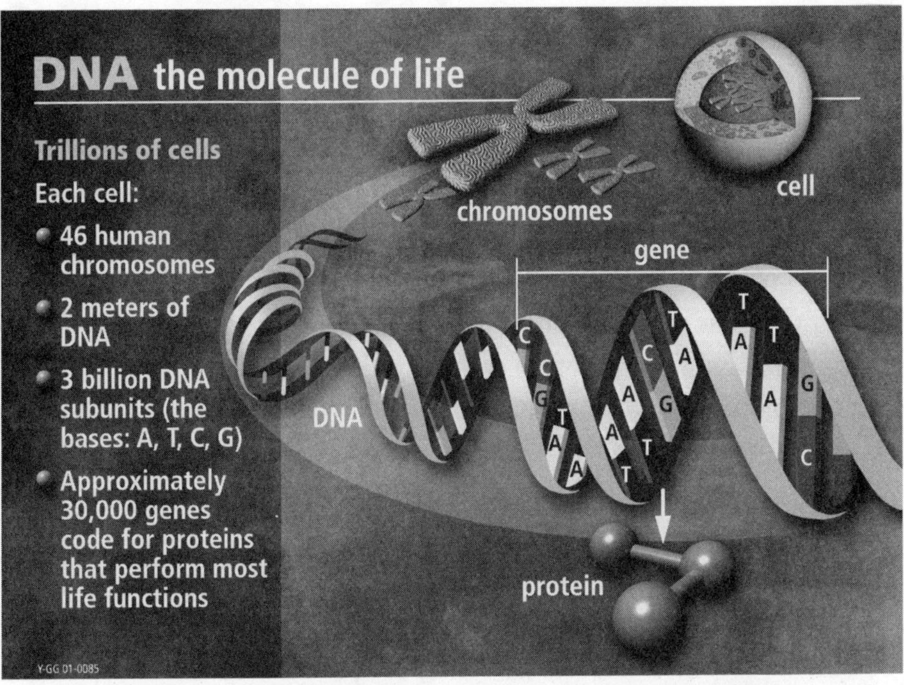

We have now successfully described "the molecule of life," but we haven't yet figured out what it all means. It's the single unifying code that *every* living thing on Earth shares. *(Courtesy of the US DOE Human Genome Project)*

medicine, I've decided against it. I mean, there are so many medical technologies out there, you could write a stack of encyclopedias about it all. In fact, other people, far more knowledgeable than I am, have. And the moment they finish them, they have to be revised. So instead of medicine, I've decided to tackle something a little more manageable.

Immortality.

Don't worry, we can handle it.

Death Avoidance

Immortality is difficult to come by. Egyptian pharaohs tried using pyramids. Roman emperors and later, the great monarchs of the Enlightenment, developed a weakness for marble statuary and oil paintings to immortalize themselves. Anything to avoid the black, humbling oblivion of death.

I suppose we all have our ways of trying to attain a kind of immortality, methods for dealing with the harsh, inescapable truth that life comes to an end. Entire religions have been erected on a foundation of death and the questions about what lies on the other side of it. For many of us we feel we can gain some modicum of immortality through our children (assuming the stress and worry of raising them don't do you in first). But I've always found that approach a little suspect. I mean, what good is immortality, if you aren't around to enjoy it?

So for some—me, for instance—these strategies just aren't enough. I mean I want the whole enchilada—long life where the end is not in sight. And I'm not talking about eking out a few extra years at the retirement home, bobbing for dentures, trying to remember where I put the Depends. I want a vigorous, healthful, butt-kicking life. On the other hand, I don't want to turn the clock back entirely and become as stupid as I was at twenty-five, for all eternity. Let me keep the experience, just give me health and vitality, and a serviceable mind to go along with it. That's the goal.

You scoff, but I am not alone. As a society we are obsessed with living longer, and youth is worshipped passionately. Just the diet section of Barnes & Noble would lead an alien to conclude that we value calories (or the lack of them) more than we value literature or even a decent line of poetry. A whole generation of baby boomers, which has been passing for forty years through the world's consumer markets like a pig through a python, is coming to terms with the inescapable fact that they are getting older and that at the end of the trail, there is only one destination—and it ain't another Rolling Stones concert. My bet is they're prepared to pretty much do whatever it takes to keep the Grim Reaper at bay.

Star Trek toyed plenty with the concept of immortality. In fact, it was one of our favorite themes. There was "Mudd's Women," eternally young and beautiful, thanks to the strange drug they took. And "Requiem for Methuselah," where we explored what a long and lonely business forever can be, especially if you happen to be the only immortal on the block. I think my personal favorite, though, is "Space Seed." That was the show that featured Ricardo Montalban as the driven and charmingly evil Khan Noonien Singh.

The episode starts out with the intrepid *Enterprise* crew coming across the *Botany Bay* drifting in space; a ship that our trusty scanners reveal is a relic from the twentieth century. Naturally we have to investigate. Once beamed aboard we find four humans in suspended animation, one of them Khan himself. Accidentally we revive Khan (by turning on the lights!).

Khan Noonien Singh (Ricardo Montalban), in the future envisioned by *Star Trek*, ruled a quarter of Earth with an iron fist from 1992 to 1996 during the Eugenics Wars. Kirk found him and his genetically enhanced crew 270 years later. In the real world the Human Genome Project was initiated in the early 1990s. Today the human genome has been mapped and the issue of eugenics is no longer a theoretical one. Will we reinvent ourselves from our DNA up? Will we use it to make death obsolete?

We then find that these are fugitives from the early 1990s, a dark period in Earth history when the great Eugenics Wars were being waged. (How far away the 1990s must have seemed in 1968! And now they are history, literally.) Anyhow, a quick check of our computer's database tells us that Khan, who is himself genetically enhanced, once ruled a quarter of Earth with a superhuman iron fist before being overthrown. After being fired from his job as a eugenic dictator, Khan, it seems, escaped into space with eighty-four hand-picked superspecimens aboard the *Botany Bay*, where, after an extended sleep, he has decided to make a comeback.

He almost succeeds too, managing briefly to take control of the *Enterprise*. Of course, my alter ego managed to outsmart and then outfight Khan (not bad considering Kirk is just working with a normal complement of genes) and maroon him and his buff followers on a woebegone planetary outpost by the name of Ceti Alpha V.

It just goes to show what can happen when you start messing with the machinery of life. You end up with a bunch of rogue superhumans who can't curb their powers or work and play well with others. Next thing your starship's in jeopardy, a sequel gets made, you lose your first officer, and end up with his alter ego directing the *next* sequel. Only in Hollyweird, only in the world of *Star Trek*.

But you have to hand it to *Star Trek*, which proved yet again to be ahead of the curve, sort of. It is true we managed to get through the 1990s without a eugenics war. I'm glad we weren't right about that. Nevertheless something fateful *did* happen in 1990 that has had a stupendous effect on the field of genetics, and will likely have far more important repercussions for the human race than Khan could ever have suspected. That year the Department of Energy and the National Institutes of Health initiated a joint research effort called the Human Genome Project. At the time there weren't very many people who really understood what this was all about, but they're finding out now. As I write this, ten years after the project began, scientists are announcing that they have successfully mapped the entire human genome. That is, they have located and plotted all of the genes within us *that make each of us possible*.

This, I have learned, could result in the death of death.

Life's Blueprint

Telling me that a battalion of scientists has undertaken mapping the human genome means about as much to me as saying, "Your hard disk is corrupted." However, I do comprehend that genes are important and that life is important and that the two are more or less connected. So it became pretty clear, pretty fast that Chip and I needed to powwow with some experts.

So, once again, off we went, up the arm of California to Walnut Creek, just outside of Oakland. We converged on a leading authority in a brand spanking new science called "genomics," Elbert Branscomb, the director of the Department of Energy's Joint Genome Institute.

The institute's offices were unassuming. We found no boiling vats of man-made pathogens or *Brave New World* cells incubating test tube babies. Nothing to indicate that the work being done here (and at several other research labs around the country) might entirely upend our view of what is natural, even what is human.

There wasn't much to see for the simple reason that, like nanotechnology, genes operate at the infinitesimally small molecular level. They're invisible. On the other hand, there was Elbert Branscomb, and he was *not* invisible. He was genelike in this sense, though—he was compact, energetic, and loaded with information.

To look at Elbert, I would have sworn he was in his fifties (actually he was about to turn sixty-six!), but once he started talking about his work, and pacing around the room and scribbling notes and pictures and formulas furiously on the wall's enormous whiteboard, it became clear this wasn't a mature physicist turned biologist, this was a twelve-year-old who had been cut loose to play with the most fun toys he could possibly imagine—the secrets of life. I've rarely seen anyone having a better time earning a living. Good thing too, because I was relying on him to explain the complexities of human genetics and I'd need all of the good humor and patience he could muster.

I did know a few basics before I walked through the door. Over the years I had picked up what a cell was, more or less, and I knew that within the nucleus of each, down there in what I now understood to be nanoland, each of them houses a long, wound up string of genes—our DNA. I vaguely recalled that our DNA is composed of two strands of chemicals that are somehow intertwined, as if you had taken a ladder, twisted it, and then squashed it down into something very, very small. (Our DNA weighs just 10 trillionths of a gram.) Generally I was aware that somehow, miraculously, these strands of chemicals and molecules contain all the information necessary to make a human being. But that was as far as I could take it. What I didn't know was what DNA was, precisely? What was a gene? What does it do? And how, exactly, does it do it?

"It's a recipe," says Branscomb, beaming, "a set of instructions for building and operating complex molecular machines. Machines that walk and breathe and talk and think, and even try to explain how the blueprint that created them in the first place works. In a sense, we are run by these genetic information systems, kind of like a robot. We are built from a single egg through all of this Rube Goldberg mechanical/biological rigmarole into a human being that walks into a bar and says, 'Hey, didn't I see you here last night?' "

"But how does it work?"

Elbert rubbed his bearded chin.

"Well, let's start with the twisted ladder. It's called a double helix—a spectacular, natural recording system. Think of it as a computer hard drive (particularly bad analogy for me, but okay), a place for recording digital information. On this invisible hard drive is written an endless string of zeroes and ones. Except our DNA doesn't use two symbols, a 0 and a 1, it uses four. We call them A, T, C, and G after the names of the nucleotides that cling to one another to form the steps of the ladder: adenine, cytosine, guanine, and thymine."

Branscomb merrily went on to explain that all of the information that constitutes you and me is encoded in these four nucleotides that cling to a long string like pop beads. This is where the entire book of life resides. The order of the A's and T's and C's and G's, the beads, spell out the instructions for our creation and operation. A separate, different kind of binding also takes place, says Elbert, but it has more to do with keeping our DNA stable and able to repair itself. This is the famous double helix, the twisted ladder-like shape of DNA. Essentially each nucleotide, or pop bead, on each string binds with a companion nucleotide like the opposite poles of a magnet. Adenine and thymine always go together and guanine and cytosine always go together. Why? Because they make a perfect fit, like two complementary puzzle pieces, or certain happy couples or some comedy teams.

"The storage device—the DNA—is really the 'dumb' part of the process. The real magic and mystery and wonder," says Elbert, "is in the machinery of proteins that float in the soup of your cell and all around the gene. The cell is like a factory. Now imagine the factory is filled with robots, amazing machines in and of themselves. These machines are proteins and they take instructions from the DNA and then go about building new versions of themselves out of material within the cell, sort of the stuff on the factory floor! They self-assemble. And once they are assembled, they do all the work that is needed to reproduce even more cells. Each protein performs its own unique role.

"It is astoundingly ordered."

Our genes are wound into tight little cassettes of about one thousand genes each called chromosomes, says Dr. Branscomb. It seems that each of us has two sets of twenty-three chromosome cassettes, one from mom, one from dad. We get this half-and-half mixture when a sperm and an egg team up to create, well . . . us. Sex, it seems, is nature's way of mixing the gene pool up to create the greatest possible diversity. The more diversity in a species, the more likely it is to survive. To survive, a species has to have its own kind around so it can keep reproducing *more* of its own kind, and that means sex is *very* important. Undoubtedly this explains why sex is so much fun. Its pleasures are your reward for doing your best to mix and match the gene pool.

Gotta love Mother Nature.

From this single sex-born cell of combined chromosomes emerges each of the five trillion cells that comprise you and me and every other human who has ever come along. Each holds within its genes everything necessary to make a complete copy of you. Or, put another way, each cell holds within it the complete recipe for making each of the other 4,999,999,999,999 cells, no matter what their function; whether they live in your pancreas or your stomach or your brain where they merrily fire away at two hundred times a second so you can read this passage and try to figure out what I am saying.

"That is an astounding amount of information to cram into something that is millions of times smaller than a pinprick," observes Elbert.

If you imagine the links in each chain of DNA as letters, the standard, and nearly weightless, human genome represents about 1,500 novels.* Or put another way, if I patiently unrav-

*When scientists recently completed the latest draft of the human genome, they found that only thirty thousand genes carried any information that was

eled all of the DNA in one cell and strung it out, it would be about two yards long.

You're thinking, Big deal, two yards, six feet. Give me a dimension I can't comprehend. Right, doesn't seem like much until you realize that we are dealing with molecules, you know the objects that are billionths of a meter? We are talking about laying down objects, one next to the other, that are millions of times smaller than a flea! And still they stack up to being as tall as Leonard Nimoy. If we took all of these strings out of every one of your five trillion cells and laid them end to end, you would have enough molecules to reach to the sun and back fifteen times. That's a lot of weenie, little specks, and a testament to the complexity that makes life possible.

One Human, Coming Up!

Finally I'm starting to get the picture, but I'm wondering, once again, how does all of this work?

"Okay," he says, "this long string of nucleotides represents volumes and volumes of recipes for transforming the stuff that floats in the air and water and food around us into living creatures. It writes these recipes using this relatively small ATGC alphabet I mentioned before.

"Being able to do this is a little bit like having a computer that

---

very important to being a human being! In other words only 1.5 percent of the long string of DNA within each of our cells seems to serve any real purpose. The rest is junk. Scientists had originally thought that at least somewhere around 5 percent of the genome would have useful information. "Bottom line is," says Elbert, "most of the 'text' in those 1,500 novels could be thrown out with no consequence, and what does matter is scattered in mostly tiny, hard-to-find fragments throughout the rest of the 1,500 novels. In other words the part of the genome that really appears to matter would fit into about twenty novels [still not anything to sniff at]. Pretty *boring* and vexing. But that's life as we find it."

is both programmed and mechanically gifted enough to assemble itself out of the material around it," says Branscomb. "The DNA uses various chemical switches and commands to assemble the material around it into the right cells at the right time. Basically the DNA tells the material: 'Okay, put together what you need to form a cell for the retina.' Or, 'You are a cell for the leg bone, you're a cell for the liver.' Somehow the cells know where they are supposed to be located and where one sort of cell stops and another group starts. This is skin and not bone. This is the iris, not the optic nerve. This is the spinal cord not the brain stem."

As a result of this miraculous chemistry we rarely end up with our kidneys in our brains or our brains in our derrieres (although some people have occasionally fooled me on this one). The whole process has an awful lot to say about the assets and liabilities we bring with us as we slide out of the birth canal. Genes determine hair color, eye color, sex and race, raw intelligence and athletic prowess. Some studies indicate a correlation between our genes and our predilection for certain diseases and addictions. A recent article in *Scientific American* by Dean Hamer (chief of gene structure and regulation at the National Cancer Institute's Laboratory of Biochemistry) speculated that genes could even account for many of our basic traits. Kirk, for example, might have been strongly coded for altruism and leadership. Spock (he is *half*-human, remember) for high intelligence and inquisitiveness. (He missed out on the humor gene, though.) Bones? Anxiety and compassion.

This isn't to say—and everyone is extremely careful about this—that we are all born preprogrammed to act out our lives according to the strict commands of our genetic code. We aren't machines, even if all of that molecular machinery is what makes us possible. But the genes that are turned on and go to work as we make our journey from fertilized egg to full-fledged human undoubtedly have a lot to say about what our lives are like between cradle and grave.

If we were pieces of architecture, I suppose you could speculate that some of us are born Tudors, some log cabins, others clapboard Victorians with deep porches, and a few of us doublewide trailers with the water hookups missing. The point is there's a lot of power in those long and winding ladders that make the cells that make us. But they aren't omnipotent.

If you are anything like me, if you think too hard about this, it makes the area right behind your eyeballs throb as if they've been whacked with a rubber mallet. But trust me when I tell you that it's important to understand these kinds of details if you want to have any hope of comprehending what can happen . . . *when* we develop the ability to start rewriting our genetic recipes.

The Human Genome Project (as well as other private sector companies that joined in the effort) represents the first concrete effort in the history of the human race to conceive variations on the themes of life. But how do you go about decoding something as strange and mysterious as an invisible string of twisted and squashed molecules? That was the next question on my mind as we sat in the conference room in Walnut Creek surrounded by whiteboards covered with Elbert's words and formulas and pictures. How in the devil did they even begin to figure out the sequence of the three billion letters that make up our DNA? I can imagine it a kind of molecular paleontology where you sift for a secret here and a fragment there hoping somehow to pull the whole complicated picture together. It must have been unbelievably complex and tedious.

"It's horrific!" says Branscomb, standing in front of a whiteboard, slashing the air with the marker in his hand. "It just drives you mad. We have to piece it all together and then . . . [laughing just a touch insanely]. Well, it makes you nuts."

And now I found myself drawn into the whole maddeningly

complex experiment too. Here we were talking about manipulating the very chemistry of life at the most basic level.

"How did you do it?" I asked.

Here's how.

Spelling "Human" One Letter at a Time

Every day at the Genome Institute (and a network of other centers around the world), researchers have been tossing nucleotides into solutions loaded with DNA for nearly a decade, and then waiting for the molecules to find their sister nucleotides. Adenine pairs up with thymine; cytosine with guanine. The Joint Genome Institute has been doing this an average of 40,000 times a day. Teams of scientists next drop what they call "terminator nucleotides" into DNA solutions. These are not Arnold Schwarzenegger cyborgs, but molecules that end or terminate a string of zipped-up nucleotides. Basically they snip the molecular strings that have formed and produced short trails of DNA. This method yields the location of *one* letter on the string—the one opposite a color-coded terminator; A if the terminator is C; T if the terminator is G, or vice-versa. Since you already know what the terminator is, you also have to know what kind of nucleotide it hooked up with, right? Next these snippets with their one color-coded nucleotide are run through a Jell-O-like solution called a "sieve matrix." As each of them percolates through the matrix they are scanned by a laser that reads the color and length of the string and codes it into a computer.

With that information researchers can then take all of the records of all of the strings, each with their terminators (A,C,T, or G), and overlay them, one on top of the other. Only then does a pattern begin to emerge. Only then can scientists say, "Ah-hah! That's the sequence of the DNA bases!"

But remember all of this work has only revealed relatively

short chains of DNA, around five hundred letters or so each. Once they were assembled then they had to be linked in proper order to other short chains of the same size, says Branscomb. Finally, after ten years of snipping and scanning hundreds of millions of times, teams at all of the laboratories have lined up the whole confounding code! As of the summer of 2000, the project had succeeded in creating a draft (not a final manuscript) of the human genetic code. More is being learned every day.

Why have so many gone to all of this painstaking trouble? Why haven't they just put the DNA under a microscope and looked at it? Because the pesky little molecules are too small!

Says Elbert, "Science currently has no instrument on Earth that can simply look at real, live DNA and just read the letters of the recipes as if it were some sort of infinitesimal microfiche. Despite all of the advances in medicine, computer science, and biotechnology, the only way to learn the order of each of the letters is to do it in this maddening, indirect way."

Humbling.

However, as you read these words, human DNA has been mapped, letter by letter, each one dropped and scanned through the sieve matrix. As you read this nearly all of the accumulated recipes are in hand, every friggin' genetic jot and tittle. And that means we are ready to cook. Now we can start editing the recipes and changing the behavior of specific genes to eliminate disease or create genetically altered tomatoes that cure cancer or custom design superbeings like Khan Noonien Singh. Maybe even wipe out death, right?

Wrong.

Why?

Because although we now have all of the letters of the recipes, we still don't have a very clear idea of what the whole mass of life-giving machinery is really telling us!

But that doesn't mean we won't figure it out.

Fascinating . . .

—Spock

20

ALIEN CODES

"Imagine you're Captain Kirk," says Dr. Elbert Branscomb, looking at me across the conference table in the white-board gabled room at the Genome Institute.

Okay, I can handle that.

"And imagine that you go to some planet where a strange, robotic creature lives."

Also not a problem.

I am sitting here gazing at Elbert as he endeavors to explain to me why, after ten years and billions of dollars, we humans do not yet understand the recipe for humanity that we have at last managed to map. I'm beginning to see his point. I mean he's using metaphors I can relate to.

"This thing that you find is thoroughly alien. You understand absolutely nothing about it and you certainly don't understand how to communicate with it. So you send Mr. Spock out to investigate and he comes back with all the computer tapes or code or whatever that run this alien robotic form of life and you say, 'Terrific, Spock. But what does it mean?' And he says, 'Well, it's a string of symbols. It's in some sort of complex pattern.'

"And you say, 'Yes, but what does it *mean?*'

"And he says, 'I have no idea, Captain.'

"That's pretty much where we sit with the human genome. We've learned a lot. We're looking at an incredibly important message, but to us it still looks like complete gibberish, totally alien. It's a secret code."

I never would have thought that a hypothetical *Star Trek* episode could act as a metaphor for the Human Genome Project!

So the good news is we have the code, we've met the creature, and it's trying to tell us something; the bad news is now we have to learn to understand its language. (Maybe we should give Alan Turing a call. Of course we'd have to clone him first.) This may *seem* outrageously frustrating, but it's a huge first step because once we do figure out the code, we can begin doing amazing things like cure diseases that result when genes send cells instructions that foul up the body's normal operations. Sometimes deadly instructions.

"Cystic fibrosis, for example," says Elbert. "It's a terrible genetic disease that slowly suffocates its victims, usually young children, because mucus accumulates in the lungs' airways. The kids can't get oxygen through the airways to the rest of the body."

Branscomb huddles down at the big conference table, as if he's sitting down to tell a juicy campfire story.

"About five years ago, after two decades of tremendous effort and many hundreds of millions of dollars, the gene responsible for the disease was finally found. This was incredibly hard even though researchers knew, roughly, where to look in the long DNA sequence for the disease's gene. They had whittled it down to a string five to ten million letters long; not bad when you're dealing with three billion. It's like looking for a particular grain of sand on a beach and then narrowing the search down to a fairly small *section* of the beach.

"Anyhow, somewhere in there they knew that they would find the culprit, and using every trick in the book they finally tracked down what they thought might be a candidate. But to be certain they had to compare the letters—the nucleotide bases—they had found in victims' DNA with the same snippet of DNA in a healthy person. Low and behold victims suffering from the disease had a different sequence of letters.

"The offender was a gene for an invisible molecular pump that sits in the membranes of lung cells and delivers chloride to our lungs. When they found it, I'm sure they must have screamed and wept and screamed and wept because for years the theory was that this terrible disease was somehow a failure of the cells of the lung to deliver the correct concentrations of chloride into the naturally occurring mucus we all have."

"Why would molecules of chloride matter?" I ask, looking at Branscomb like an eight-year-old kid being told a whopper.

"See, the solubility of the mucus—basically how thick it should be—is governed by the concentrations of tiny chloride ions (electrically charged atoms), and when the concentration is wrong, the mucus grows stiff and will not work as the cleaning mechanism it is meant to be for the lungs. It won't flow and it makes breathing extremely difficult."

This is really amazing to me. That the body is constantly—at blurring speed—manufacturing millions upon millions of tiny machines that engineer everything from our thoughts to the breaths we take to stay alive. It's miraculous! It turns out that when the genes that control the injection of chloride ions in our lungs work the way they should, amino acids—small groups of molecules—literally fold up and form a pump that deliver the chloride ions down something like a drain. At the bottom of this channel, there is a flap, a kind of valve, an inconceivably tiny, minuscule, invisible valve, which controls the flow of chloride, one ion at a time, into the cells of the lungs.

When the whole system works, when our DNA gives the cells and molecules the correct instructions, we all breathe easier. But for people with cystic fibrosis, Elbert explained with an intensity bordering on ferocity, a part of the protein in the DNA that controls the operation of the flap malfunctions. The valve doesn't work correctly; it doesn't let enough chloride ions into the lung cells, and without enough chloride, the mucus in the lungs becomes thick and hard and blocks the airways.

So, now that they had found the bad guy, did they do him in?

"Yeah, but to do that they had to get a new, correct set of instructions into those lung cells."

"How do they do that?"

"Well," says Branscomb, "there have been experiments in lab cultures and in certain animals that show that when you get the right proteins in a cell, it begins to absorb the correct proteins and the DNA begins to make new versions of itself. In other words you get the right letters in the right places in the alien code and they fall into place! The faulty DNA *wants* to do the right thing, but it needs help. The recipe is rewritten by getting the right proteins into the cells which, in turn, deliver the right instructions. The bad guy disappears!

"But that's experimental. With luck, though, after all of this pain and agony, we will soon be able to inject these "good" proteins into human victims, the children that have this affliction, and we will have conquered our first genetic disease!"

Phew! A happy ending.

So what does any of this have to do with immortality? I'm getting to that.

The cystic fibrosis story illustrates the big hope that drives the Human Genome Project: the eradication of genetic disease. That's certainly one way to lengthen our lives and start sniffing around the impossible possibilities of life without death.

There are long lists of sicknesses that hobble some of us

because our DNA misfires: diabetes, schizophrenia, sickle cell anemia, high blood pressure, Alzheimer's, predilections for certain forms of cancer. The list of deranged strings of DNA that can annihilate us is long and vile. Now we may no longer have to take this abuse. We may be able to soon go in and tinker with the machinery of life (once we figure out what it's trying to tell us) and voilà! We're running like a Swiss watch.

Organ-I-zation

However, not every disease that kills is the result of DNA codes gone AWOL. The experts will tell you that there are plenty of other nongenetic killers out there, like heart disease or tuberculosis or AIDS. The damage that these afflictions do is not the result of inborn molecular snafus, but the environment, or our own lack of sensible living. We damage our hearts by smoking or eating too much or we catch a virus or bacterial infection that does lethal damage to our kidneys or lungs or some other organ that our bodies like to have operating at full capacity to keep the engine running.

As it turns out, the secrets that lie within that long, opaque string of nucleotides that are arranged just so can also help us with these problems too. But how, if they aren't directly related to our DNA?

Well, for one thing, manipulating DNA may make it possible to grow brand-new organs, just for you, although it's unlikely, you will be going online any time soon to order up a new kidney or an improved pancreas.

One of the really scary issues floating around the whole idea of genetic engineering is human cloning. It's a fear that goes way back to Aldous Huxley's science fiction tour de force *Brave New World*, and there are plenty of good reasons why it's scary. For those of you who may have been stranded on a desert island and missed a few issues of *Time* magazine, cloning is a process by which you grow a completely new version of you from a single cell harvested from your original self. On October 13 of 2001, Advanced Cell Technology cloned a human embryo. A lot of ink has been spilled discussing the ethics of human cloning, but personally I'm not sure that we have to worry about making complete copies of ourselves because I'm leaning toward an alternative: selectively cloning just those parts of us that are breaking down.

Strange? Yes. Will we start seeing Trak Human Body Parts stores springing up in mini-malls all around the world any time soon? I hope not. But then, maybe it's not all *that* strange when you consider that each year surgeons replace the vital organs of thousands of people with organs from *other* people who have already died. (We are now looking at using pig organs as replacements!) These operations are costly and rife with trouble. I have been told, for example, that your body apparently doesn't care for having someone else's organ in it. It instantly recognizes it as foreign and immediately tries to boot it out, as if it were a massively lethal bacterium, or a Democrat in Orange County. All of the forces of your immune system are marshaled to destroy it. And when you need that organ to survive, that's a problem.

To overcome organ rejection, medicines have been developed that suppress the immune system so that the body will stop destroying the very organ that is saving it. But then, it turns out, *that* leaves you with a severely weakened immune system which makes you susceptible to countless *other* diseases. Add to this the tragic fact that far more people need organ transplants than get them. I looked it up. In 1997 40,000 people were on the waiting list for a new heart. Twenty-three hundred got one.

All of this makes you start wondering if there isn't a better way.

I'm just speculating here, because what do I know, but as far as I can see organ cloning would eliminate all of these problems. When an organ has been grown from your very own cells, it won't reject it, which also means you can preserve a perfectly working immune system after the new part is installed. Nor should supplies be a problem. After all, cells are far more plentiful than fully developed hearts or kidneys. Add to this that no one needs to die in order to provide the replacement part in the first place and organ cloning starts looking less insane than it did four paragraphs ago.

Does this make me feel skittish? Yes! But not as skittish as putting a pig heart inside of me!

Some of the scientists who think about these things point out that more than organs could be grown in the future. Using techniques developed at the Harvard Medical School and Massachusetts Institute of Technology, researchers are cultivating artificial skin, ears, even noses that can be implanted to replace damaged originals. New cartilage may soon be grown right inside of tired and damaged joints, giving them new vitality and their owners new strength. Think of what this could do for football! Experiments with special proteins that can grow new bone and blood vessels are also underway.

Some scientists speculate that the next step is to cultivate neo-organs right on site; inside your own body. Believe it or not I came across this reading a back issue of *Scientific American* (are you impressed?). Weird as this seemed to me, I gave it more thought. There are some advantages, I said to myself, feeling to see if my kidneys and heart seemed to be operating okay. After all, why grow a spare kidney in the lab when your body has all the right enzymes and proteins and genetic wisdom to grow your new part right where it belongs, just like it did when you were in the womb? (Of course this could get to be a problem if you wanted to double up on *all* of your major organs.)

The approach (still in the experimental stage) is being pioneered at the University of Massachusetts Medical School and Harvard Medical School's Children's Hospital and will allow doctors to transfer the appropriate cells to the desired site. Once the correct ingredients are located in the right spot, the body pretty much takes it from there and grows a new organ, which can later be transplanted to replace the old one. At least that's the theory.

Gene therapy, the kind we discussed with Elbert Branscomb, can kick these efforts to another level because it is our DNA that seems to know when to command various cells to begin creating the right body parts in the right location at the right time. The more you understand about the operation of genes, the more successfully you can pull off tricks like this.

Where will all of this lead? Well, maybe not immortality exactly, but almost certainly to a longer, healthier life (assuming you don't go down in a plane crash or get clobbered by a bus), and that might be just as good. Hey, at the very least, I may yet be able to return as Captain James T. Kirk. This time in *Star Trek XIV: The Resurrection*.

The Deadly Years

There are other advantages to life-extending techniques like these. It's true, a home-grown kidney won't, on its own, guarantee instant immortality, but could buy time enough to locate the brass ring—a true fountain of youth that could make extremely long lives possible. After all, given the pace of advancements in medicine—which blisters my brain—and given the fact that we have just mapped the human genome, isn't it possible that we might all get our hands on an extra ten or fifteen years of good living pretty soon? And if we do that, isn't it possible that additional advancements during those fifteen years will buy us yet *more* time, until we can eventually eliminate all of these interim steps, just cut right to the chase and wipe out old age itself, the ultimate genetic disease?

Maybe.

Remember a *Star Trek* episode entitled "The Deadly Years"? In that episode Spock, Scotty, McCoy, and Kirk all aged right before your very eyes at a rate of thirty years per day.

I can't say the science was very solid for this particular episode, but the premise was great. After visiting an experimental colony on planet Gamma Hydra IV (who comes up with these planetary names anyhow?), our landing party, which also included Chekov, find several members of the colony looking, well, like hell. I mean really ancient. So we beam them back on board to try to figure out what's going on. No sooner do we return than Scotty, Spock, Bones, and Kirk start to age faster than air traffic controllers at O'Hare Airport. Chekov, however, remains his same youthful self.

After a commercial break we burn the few aging synapses we have left and figure out that Gamma Hydra IV had passed through the tail of a comet loaded with low-level radiation. That, we deduce, is the cause of the accelerated aging (told you the science was weak). Anyhow, a few more pages into the script and a crotchety and even more cranky McCoy figures out that the one emotion that the rest of us didn't experience, but that Chekov did, when on the planet was fear. You see, back when we had landed on the planet, Chekov had had the bejesus scared out of him when he found the first thoroughly aged (and dead) colonist. The fear had basically given him a jolt of adrenaline, and that, Bones concluded, must be the antidote. So we all received megadoses of adrenaline (I do this every morning when I have my ritual machiatto), and instantly returned to our youthful selves.

Once again all was right with the universe.

I bring this up because it serves as a dramatic reminder:

Aging is a royal pain in the astronomical calendar. Just as we couldn't understand why we were aging so rapidly in "The Deadly Years," we have no clue as to why we age during the normal course of our lives. I'm pretty sure that injecting people with a bunch of adrenaline isn't going to fix the problem (though we might get the laundry done a little faster), but genetic engineering could stop the cellular unraveling that ultimately sends us all to the grave.

Outfoxing Old Age

Scientists tell me that we probably age for scores of reasons, but one theory is that nature doesn't generally have much use for old members of any species. Let's face facts, the primary reason a member of a species exists at all is to ensure the continued survival of the breed. That little job is accomplished by having sex! So we reach puberty, gain the ability to procreate, do our job and, having served our purpose, begin our physical decline. All of the evidence tells scientists that after we come through our teens our bodies simply start to fall apart, cell by cell, organ by organ. Our skin loses its resiliency and ability to hold moisture, our joints creak, bones weaken, vital organs start to wear out. In an earlier era, those of us no longer fleet of foot and clear of eye would quickly wind up dinner for some big cat roaming the savanna.

Most creatures seem willing enough to accept Nature's verdict on this, but being conscious, we humans have developed a desire to hang around, even if we *have* outlived our original purpose. And this explains why we've chucked billions of dollars into exercise books and diet books and cosmetics and vitamins and science and medicine all hell-bent on outfoxing age, disease, and death.

Our desire to extend life, however, still hasn't helped us

explain the *mechanism* for aging. Why do we age at all? Cells obviously have this wonderful ability to replicate (thanks to our DNA). We are, each of us, the living, breathing proof. And through our adolescence, our allotment of five trillion cells seems to replace themselves perfectly; deterioration is virtually zip. All cells die, but, I must ask, if they can replace themselves with perfectly operating successors when we're thirteen, then why not when we're thirty or sixty or ninety, or, for that matter, nine hundred? Why must we be subjected to the fate of Gamma Hydra's colonists (albeit at a considerably slower rate) when nature seems capable of such miraculous regeneration?

I did some research on this and it seems that there are a couple of candidates. Among them something called telomeres—caps of DNA that protect the ends of our chromosomes. We all have them, but over time, as cells reproduce, telomeres get shorter and shorter, and when they get short enough, they seem to signal the cell to stop dividing. At least this is what researchers see in laboratory culture dishes. What we don't know is if this is what happens in our bodies. Apparently if you take the cells of someone who is elderly, and put them in a lab petrie dish, they divide just fine. So there's not a clear answer.

It may be that telomeres, once they begin to shrivel, signal cells in specific parts of our bodies where cells need to rapidly regenerate to stop reproducing. These would be places like our immune system or our skin and bones. And if our immune system starts to break down because we have battalions of atrophied telomeres sloshing around, who knows what additional damage may result? There may be a kind of domino effect.

So, perhaps, once we decipher at least some of the genetic code, researchers will understand how we might change the behavior of telomeres; get them to rejuvenate themselves and keep our skin glowing, our immune system humming, and our bones strong. That alone would keep us all from turning into the old codgers that walked the *Enterprise* bridge in "The Deadly Years."

Cell Pillaging Rogues

There are other sources of aging (not counting teenagers, traffic, and endless Hollywood meetings). Almost all of us have heard about "free radicals," but what are they and do they need a good lawyer? In fact free radicals are something researchers call super-oxides, kind of a rogue gang of oxygen molecules that are created as various cells break down oxygen inside our bodies. You might wonder why oxygen does any damage. After all we need it to breathe, right? Yes, but it's also apparently a pretty toxic gas that tends, when it clots into these gangs, to damage fatty compounds that the body requires. They can even inflict pain and suffering on our DNA itself.

To be useful oxygen needs to work with other molecules, but left to its own devices it's like a molecular terrorist blowing up nano-sized sections of our cells. Over time the damage adds up. Some animals (and probably some people) handle free radicals better than others do. Those that do (pigeons, for example) live longer. (Please tell me that pigeons are *not* the key to immortality.) The point is that the secret to eternal youth is obviously not located at some bubbling fountain near a lush grove of palm trees in Florida as the Spanish explorer Ponce de Leon had long ago hoped. (All *his* search led to was a record number of retirement communities in Dade County. More like a Fountain of Oldth!)

No, I'm afraid all of the evidence indicates that the fountain of youth resides somewhere within us, if it resides anywhere at all; way down in nanoland where the life-giving magic our DNA has does its work. So if we have any hope of seriously extending our lives, unlocking the genetic code certainly can't hurt the effort. All of which brings us back to the work of Elbert Branscomb and the other legions of scientists now working to understand what those molecular codes within us actually mean. Maybe it's telomeres, maybe it's busting up gangs of free radicals, maybe it's coming up with ways to change the way certain genes react to hormones that resemble insulin (this works for roundworms!). Whatever it is, it's not obvious and it's not in a fountain of water. And it's not in a shot of adrenaline. But don't give up hope. Science still has a few experiments up its sleeve, as you'll see if you read on.

The best is yet to come, and won't that be fine?
You think you've seen the sun,
but you ain't seen it shine . . .

—The Best Is Yet to Come
Lyrics by Carolyn Leigh

Don't leave him in the hands of twentieth century medicine!

—McCoy to Kirk
Star Trek IV: The Voyage Home

21

SUSPENDED ANIMATION 101

If you have even the slimmest experience with science fiction, you've very likely come across the term "suspended animation." It's a pretty time-worn old saw whereby humans essentially go into a deep sleep, a hibernation of sorts, slow their bodily functions down so much that consciousness is obliterated and their metabolism grinds to a near halt. It's a neat trick and another way to deal with the enormous distances the universe imposes upon us.

Without a trick like this, you would simply die of old age long before you reach your destination, and that wouldn't be any fun. So you suspend the animation of your bodily functions and bring your bodily clock to a halt so you can be around when you finally get to where you are headed.

We didn't use suspended animation much in *Star Trek* because we had those fabulous warp engines, but over the years it's shown up in everything from *2001: A Space Odyssey* to *Aliens*, not to mention countless science fiction novels.

One time we did use it in "Space Seed," the same show that featured Khan and his gang on the *Botany Bay*. When McCoy and Spock and Kirk beamed aboard we found that the bodies of Khan and his cohorts weren't running any faster than traffic on the Ventura Freeway. Heartbeats, respiration, body temperature—all of their metabolisms were just crawling along. It was a nice little dramatic trick that made for an interesting story, but not anything we have much use for in real life. Why would we want to slow down our metabolisms so much that life comes virtually to an end? Where's the fun in that?

Well . . . think about it. It may be another great way to sidestep the grave. Let's say you're dying of old age or a lethal disease or a mortal injury. So you postpone death by going into suspended animation, and then have yourself revived *after* science and medicine have come up with some way to cure your disease or repair your injury or undo the ravages of old age. Crazy, right? Well, not really. Something like suspended animation already exists and it's gaining more credibility every day. It goes by the strange name "biostasis."

Under the current rules of life, we all die. (Hate that.) The Grim Reaper is an unrelenting adversary (assuming you see death as an adversary—some folks don't), and he always wins. Always. But maybe, the supporters of biostasis say, he doesn't have to. Maybe there's another way to look at what dying means exactly.

The reason we die is because our bodies stop working. I know this seems obvious, but I have a point I'm trying to make here. Physically whenever your heart shuts down, for whatever reason, any biologist can tell you that all of the 5 trillion cells that constitute you are left to fend for themselves; their support system is gone. They don't do so well that way. The heart provides the body's cells with fresh supplies and nutrients of all kinds. Left on their own, they simply don't have very many resources, and if those resources aren't replenished, toxins build up and the cells stop working. Without any help, they can survive on their own from four to eight minutes. After that, it's curtains.

There are occasional exceptions. People have been successfully resuscitated up to an hour after they apparently drowned in icy water, for example, because the cold water slows the metabolism, the cells don't use up their reserves so rapidly, and they survive longer. But no matter what, after your heart stops beating, cellular damage, sooner or later, becomes irreversible and death inevitable.

Freeze-Dried You

This is when, many will argue, all sorts of metaphysical events also begin to unfold. Our souls, our consciousness, our essence leave our nonfunctional bodies and migrate elsewhere. But that's a different book.

Most doctors today would agree that we die when our hearts have stopped pumping. Within four to eight minutes the damage is irreversible and our key organs are out of commission, most notably our brains. Biostasis experts, however, believe there may be ways to extend the grace period during which our cells can survive. How?

By freezing us.

For years scientists have known that all kinds of cells can be frozen and then rejuvenated when thawed. In Siberia, for example, in the 1980s scientists drilled cores out of the permafrost and found microbes that were 100,000 years old, frozen solid. When they warmed them up and put them in a solution of water, they were good as new. The *Journal of the American Medical Association* has even published articles about patients suffering from hypothermia (critically low body temperature) who had heart attacks. For some time after the heart attack there was absolutely no brain activity, yet when the patient's blood was warmed and primed with oxygen and glucose, life resumed without any signs of brain damage.

So there's obviously *something* to this freezing thing, but the $64,000 question is: Just how far can you lengthen the "grace period" by cooling the body and all of its cells? And can you really make it work for humans? Given the territory we are exploring, it was clearly time to consult another expert.

Waiting for McCoy, and Other Experiments

You'll never guess who happens to be one of the world's leading authorities on this. None other than the irrepressible Ralph Merkle, Ph.D., Eric Drexler's cohort in the field of nanotechnology. (There's a connection between biostasis and nanotechnology that you will soon see.) Ralph's qualifications in this subject meant it was time to have another luncheon meeting with his Merkleness so that we could take up the issue of immortality while we also fed our faces. So we met up at a pleasant French restaurant in Silicon Valley where we tackled the issue of eternal life over a meal that surely raised our cholesterol to dangerously high levels.

"Cryonics, or biostasis," says Ralph, "is really a pretty simple proposal. It goes like this: Current medical technology is not able to keep us alive let alone healthy well beyond our current life expectancy. Future medical technologies, on the other hand, look like they might be able to keep us both alive *and* healthy for very long periods of time, unless we suffer some truly remarkable damage like being clobbered by a bus or something unpleasant like that. The problem of course is that today's medical technology is all we've got and the future medical technology won't be available until . . . well, the future. So the trick is, how do we survive long enough to reap the benefits that future medical technology may be able to offer? The answer, at least the cryonic answer, is: freeze a person who is not able to be kept alive today and store them at the temperature of liquid nitrogen (77 degrees Kelvin) until tomorrow."

For those of you who can't do the Kelvin to Fahrenheit math, your suspicions are correct. Seventy-seven degrees Kelvin is very cold. That's cold enough to pretty much bring all chemical reactions in your body to an utter, screeching halt. At 77 degrees Kelvin (–320 degrees Fahrenheit!) hardly a molecule is moving, at least within your body.*

"And you can remain in that state quite literally for centuries or even thousands of years essentially unchanged," says Ralph.

And this holds death at bay?

"Think of it as an experiment. (1) You select N subjects; (2) you freeze them; (3) you wait a hundred years; (4) you see if the technology of 2100 can revive them.

"In a scientific experiment you call those that the experiment is being performed *upon* the experimental group. Those that are unchanged, or *not* experimented upon, you call the control group."

Ralph is willing to admit that this is a pretty severe experiment. After all, you have to wait a hundred years for the results. And you go into the experiment blind as Geordi La Forge without his VISOR. There's absolutely no way of knowing right now if biostasis works; no proof that you can even revive a mouse or a dog or a stone-hearted New York lawyer after he's been frozen.

On the other hand, consider the alternative.

"You have to ask yourself this: Would you rather be in the

---

*"Actually," says Ralph, "to be picky, they are moving just a little bit; wiggling really, but not going anywhere."

control group or the experimental group," says Ralph, gazing over his iced tea and looking like the Cheshire Cat. "I already know what happens to the control group. They die, every one of them. On the other hand, the jury is still out when it comes to the experimental group. Personally, I'd rather be in the experimental group. What's the downside?"

He's got a point. The only time you are going to be frozen is when you are already a goner. Either way the worst thing that happens is that you die. Biostasis at least gives you a shot at resurrection. (I just had a wild thought. What if by 2100 pretty much every conscious being on Earth is a robot and none of them give a damn whether you're revived or not? Have to check with Hans Moravec on that!)

Stop Death in Its Tracks

Ralph is so serious about participating in the experiment that he has signed on as a customer with a nonprofit organization called the Alcor Life Extension Foundation. Alcor specializes in freezing its clients as quickly as possible after they have been pronounced (legally) dead. To ensure there are no mix-ups, every member wears a loose stainless-steel bracelet on his or her left wrist with precise instructions to whomever happens to be around when they exhale their last (at least last for a while) breath.

Merkle's reads as follows:

MED. HX. CALL 24 HRS.
800-367-2228 OR
COLLECT 480-922-9013
IN CASE/DEATH SEE REVERSE
FOR BIOSTASIS PROTOCOL
REWARD A-1173

"Okay," I say around a mouthful of grilled veal chop (braised with fennel and rosemary, served over cannelloni beans), "so one day, by current definitions of death, let's say I kick the bucket. The great Starfleet admiral in the sky has called me to my final assignment, except I'm not so sure I'm ready to comply. Instead I have my Alcor bracelet on and Alcor gets a call. Then what?"

"They take over," says Ralph.

In fact, if possible, he explains, the first Alcor team has been on standby waiting for my fateful final breath. (In other words if I've been "dying" slowly as opposed to being trampled by one of my horses, Alcor has been waiting.) The first team consists of four medical technicians, one of them an emergency medical technician (EMT) with advanced training in what Ralph calls "vascular access." This apparently means they know how to pump things into my veins. In this particular case, the EMT pumps what's called a "washout" solution into me, the same solution used to cleanse organs before they are transplanted.

"This gets rid of the bad, unreplenished blood in your body and stops the cell damage," Ralph explains as he pops some gnocchi fra diablo sautéed in a spicy red sauce into his mouth. (I can almost *hear* our hearts clogging up.) "It's been cooled to a temperature of 5 degrees Celsius (that's 41 degrees Fahrenheit). This is usually done under the supervision of an Alcor physician. They like to begin the procedure as soon as possible after your heart stops."

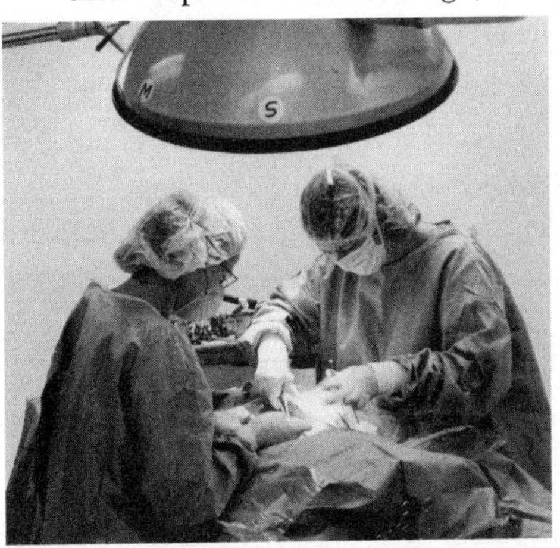
In stage one, to escape death, your body temperature is dropped way down. . . . *(Alcor Corporation)*

The goal, says Ralph, as

Then you're placed in storage until science has found a way to resurrect you. You'll be steeped in liquid nitrogen and dropped to a temperature near absolute zero. But you won't notice. *(Alcor Corporation)*

we start eyeing the dessert menu, is to have you completely refrigerated within two hours of death. The sooner the better. In fact, because more and more cells are damaged—especially in the noggin—as toxins build up at normal temperature, the faster this gets done the better your chances of making a proper recovery sometime in the future (if there ever *is* a recovery).

Next your chilled self is whisked off to Alcor's facilities in Scottsdale, Arizona (odd place considering cold temperatures are so necessary, isn't it?). Once there another team goes to work. The first thing that happens is that you're pumped full of, you'll never guess—antifreeze. You know the stuff that keeps your car radiator from becoming a block of ice in the winter, and the fluid that is also used to chill and preserve organs for transplant operations. ("There's a slight difference between the automotive and medical versions, but not much," says Ralph.) This drops the temperature of your body, or you (depending on how you like to look at it) to minus 320 degrees Fahrenheit, the level at which nitrogen changes from gas to liquid.

Everything under your skin has now come to a skidding halt.

You are then placed in a large stainless-steel dewar, a kind of oversized thermos bottle. There you are kept thoroughly chilled with regular helpings of liquid nitrogen; consigned, not to your Maker, but to the future. The question now is: Can you be revived? Are you frozen solid, beyond saving, or have you just undergone the most intricate form of first aid imaginable?

Nanotechnology Revisited

Now enter, once again, nanotechnology, which is either one of the great transforming technological breakthroughs in all of human history, or one of its most elaborate flea circuses.

Ralph reminds me of what nanotechnology can theoretically deliver: Robots that are one hundred billionths of a meter from tip to toe, give or take an atom. They can work inside the body, and be programmed to do very precise work at a very precise scale, like get your brain up and running again.

Just for the sake of simplicity, let's say that I died of a heart attack induced by an overly zealous Vulcan pinch in the neck. I wasn't trampled by my horse, or my legions of fans. It's just that the old ticker gave out.

Someone dials 911 (how ironic!), the paramedics arrive, look me over and pronounce me a goner. Then they see the stainless-steel bracelet on my left wrist. A phone call is made to Arizona and now the Alcor team moves in. I am immediately refrigerated, rushed off to Scottsdale and placed in my very own, very cold thermos where all chemical traffic in my brain and body comes to an absolute standstill. Or to use familiar science fiction terminology: I am placed in a deep state of suspended animation.

Now it's seventy-five years later and the world is swimming in nanotechnologies of all stripes, including the kind that can be used to defrost and revive me. Why are nanotechnologies even needed? Why can't we just turn the thermometer up and defrost me?

"Well," explains Ralph, "even though you haven't been hammered by a drunk Klingon, a lot of other invisible damage has been done. When you're frozen, your cells shrink because the water in them (and we are mostly water) is drawn out of your cells and freezes all around them. And because water expands, it damages the cells even more than they are already damaged. There may be ripped axons or hemorrhaged capillaries; or myelin, the material that covers your nerves, may unravel. We may soon be able to reduce a lot of this damage with a process called vitrification, but right now we still have to worry about it." (See "Immortality Bulletin.")

Sound daunting? Damn right, but remember we will have entered the age of nanotechnology, *and* we are seventy-five years smarter. Given that we have learned an awful lot in the field of medicine over the past seventy-five years and given that we will have learned incomprehensibly more just in the next decades, it is not insane to believe that we may well be capable of reviving a frozen human being.

Here's how.

Thawbones—Reheating Your Way to Life

The first goal, Ralph explains, is to repair the cellular damage caused when oxygen first stopped flowing through my body after my heart gave out. Even though I was cooled down soon after I stopped breathing, all of the nasty toxins in my cells have to be eliminated or my memory upon resurrection might be worse than it already is. I can't afford that.

IMMORTALITY BULLETIN: DEATH BE EVEN LESS PROUD

Since I wrote this chapter, Ralph Merkle tells me that there've been a few breakthroughs in biostasis that could significantly improve your chances for living a life longer than you originally anticipated, assuming, that is, that you're willing to sign up for the big chill. Not only that, it could also mean being resuscitated sooner so you can enjoy more blue-light specials at Kmart! The process is called vitrification, a ten-dollar word that basically means scientists have figured out a way to eliminate that nasty ice damage that scrunches your cells when they're frozen.

"With vitrification," says Ralph, "the freezing procedure starts shortly after your heart has stopped, so there should be very little harm done to your cells."

What vitrification does, essentially, is stop biological time . . . cold. But because of the way it does it no ice crystals form during the cooling process. In the old method (being used when Ralph and I first talked) physicians mixed water with anitfreeze chemicals (cryoprotectants), pumped the solution into your body, and then dropped your body temperature to nearly absolute zero. The problem is that because water was mixed with the antifreeze, and because your body temperature can be dropped only so fast, millions of tiny ice crystals formed and tore cells apart. These aren't huge tears, but tears nevertheless, that would one way or another have to be repaired in the future. If they aren't, you might come back a little damaged, like that Michael Keaton character in *Multiplicity* who saved pizza in his wallet and wore rubber boots on his head.

Anyhow the only way to fix all of that molecular damage would be to send in battalions of nanomachines designed to repair you cell by cell. That's a tall order even for nanotechnology. And even if the technology *is* someday capable of tackling that kind of job (Ralph and Eric assure me it will be), it will take time to develop it and make it available. So anything that can reduce the damage of the freezing process in the first place means that you could be revived sooner, and after all that's the idea, right? You want to get past the freezing part and on with the living part as quickly as possible.

Vitrification has already been used successfully to freeze very small parts of the anatomy like embryos, ova, ovaries, skin, pancreatic islets, even blood vessels for transplant operations, says Ralph. But these are all tiny and they can be frozen rapidly with less antifreeze. And rapid freezing is important if you don't want ice crystals forming. If you freeze something rapidly enough the water molecules simply don't have time to join together to form damaging ice crystals.

Scientists have been struggling with better ways to freeze large objects, like your entire body (as opposed to small parts of it), or even medium-sized parts of it, like your brain. It's been a problem because they take longer to chill. (A Popsicle freezes much faster than a gallon of water.) To reduce the formation of ice crystals in larger objects physicians would have to increase the amount of antifreeze in the solution that is pumped into your body after your heart stops. And that, in turn, means that you would be thoroughly pickled in cryoprotectants. Which would be fine except that the reason physicians currently mix as much water as they do with the antifreeze is because the antifreeze itself is toxic and in extremely high concentrations the pickling might irreversibly damage your cells in ways even worse than ice crystals.

But now, says Ralph, some of these problems have been solved. Three breakthroughs now make it possible to vitrify more quickly and with less toxic antifreeze.

"First," he says, "scientists have discovered new ways to mix cryoprotectants in a way that makes them less poisonous than they have been. Second, new 'ice blocking' chemicals have been developed that significantly reduce the amount of antifreeze that has to be used. That reduces the toxicity even more. Third, researchers have found a way to chill cryopatients up to ten times faster than before." (Note to the squeamish: this only works for cryopatients who have chosen to be frozen from the neck up. Those, in other words, who have decided to preserve only their brains, not their bodies. A new body, they figure, can be added on later, grown from DNA preserved when they are frozen. This is one of many places in the future where cloning will meet nanotechnology. Ralph says they are still working on ways to vitrify the whole body for those that feel a special connection to it.)

What these breakthroughs mean is that rather than being frozen and damaged, the vitrification will encase every one of your cells in a kind of solid glass that utterly stops all biological activity in its tracks. (Vitrify means to convert into glass or a glassy substance.) The result: every cell and molecule will have been encased in amber, like a primordial insect. No ice formation, no damage.

Does this mean no need for nanotechnology too? Not likely, says Ralph. Nanomedicine may still be needed to repair "cracking" that can take place in the glass that encases your body. And there is still the damage that might have set in before the cooling process began when your cells were starved for blood, oxygen, and nutrients. And finally there is the damage that stopped your heart in the first place; the stroke or heart attack or blow to the head that "killed" you. All of those have to be repaired.

The good news about vitrification, says Ralph, is that it almost guarantees that you'll be up and about a lot sooner than you might have been under the old method, shortening the suspension of your animation. Who knows, you may get a chance to see Britney Spears do her farewell concert at age seventy-five.

"Success looks pretty likely," Ralph says. "I mean, once again, I like it a lot better than the alternative."

Second, we have to eliminate all of that frozen water that leaked out of my cells and froze in the spaces between them. That's got to go or when I am thawed all sorts of strange molecular cocktails will appear and nobody knows what the results will be.

"So," I ask Ralph, immediately after ordering some crème brûlée, "how do we handle these problems?"

"Simple. Over the course of several days billions of nanomachines are injected into your still chilly body and brain."

"Oh," I say.

"And once there," he continues as if this is as normal as brushing your teeth, "they break up the chunks of ice that have formed between your neurons—we're concentrating on the brain here—like a battalion of impossibly small jackhammers."

Next, as they break up those water molecules, the nanodoctor injects another corps of machines that evacuate the ice from my brain using a kind of molecular bucket brigade. This might take months to accomplish, but that's okay. I've been waiting seventy-five years, and truthfully, I've been so out of it I don't even know that much time has passed. In fact, I don't know that *any* time has passed. So, take your time, doc. And get it right!

Once all of the ice has been disposed of, sixty percent of the space in my head is now wide open (remember most of the brain is water, not neurons. In Chip's case it might be a little more, but don't tell him I said that). Given that all of this space is available to work with, the doctors of the future now send another army of nanomachines into my cranium and use it to set up a communications and transportation system.

"Basically," Ralph explains, gulping down some lemonade and a bite of chocolate ganache cake, "they create a kind of molecular metabrain that busily maps the terrain looking for lesions and potholes and blown out synapses and axons. Anything they find, they map and report on. At the end of the whole process—which is also very time-consuming—a wiring diagram of your brain, good parts and bad, is created by the metabrain."

"Fascinating," I say, scraping up the last of my crème brûlée and sending my cholesterol count into orbit.

Now, says Ralph, the next wave of nanobots goes in to do the repairs—molecule by molecule—that the metabrain ordered based on its earlier reconnaissance. Axons are rebuilt, membranes are patched, and enzymes are put back where they were before death. Orange barrels and cones are distributed along the neural highways and some nanobots lean on their jackhammers and tell dirty nanojokes to one another. Whatever the case, each group of nanomachines is programmed according to the wiring diagram that the metabrain generated to go in like a SWAT team and do what has to be done. And they do it at varying temperatures as my body is slowly warmed.

"Now," Ralph goes on, "as you continue to warm up, it's time to get the water back into the neurons that the freezing process shriveled seventy-five years earlier. Otherwise, you end up with a lethal cryonic hangover."

To accomplish this, whatever antifreeze is still among my 100 billion neurons is now pumped out and salt water is pumped in. There are still a few million repairbots inside roaming around, doing their work as the brain reaches a temperature of 32 degrees F, the freezing point.

Now the trick here, says Ralph, is to warm the brain slowly, and check the repairs and functions step-by-step to reduce the chances of damage. Once my brain reaches 98 degrees F, it starts operating on its own in very basic ways, but I'm still not conscious.

Now, you may wonder, while all of this has been going on, while squads of nanobots have been rebuilding my central nervous system, what's happening with the rest of me. After all, it was a bad heart that got me in this mess in the first place. Well, thanks to all of the advances that have been made in genetic engineering and nanotechnology, I've been getting seriously, cellularly buffed.

My heart has not only been repaired and my vascular system thoroughly vacuumed, but my DNA has been scanned and rewired to eliminate any genetic snafus that may have been discovered. I've also had a lot of muscle rebuilt and flab eliminated. My skin has been rejuvenated, all cell by cell, molecule by molecule. Vain you say? Maybe. But who among us wouldn't rather return with the bodies we walked around in when we were, say, nineteen or twenty-seven or thirty-eight, than the ones we nearly took, gasping, to the grave (or in this case to the refrigerator)?

It's not as though I have changed myself. I've just turned back the clock. It's a practical issue as much as anything. If you were to leave this veil of tears sometime in the future and you were, if not doddering, at least not at the top of your physical game, you wouldn't want to return with a freshly minted brain consigned to your old body, would you? I wouldn't. If, in the future, I am faced with living a much longer life, I'd just as soon do it in a body that doesn't slow me down (and has had all the crème brûlée cleared out). Otherwise why did I ever put the Alcor bracelet on in the first place?

The underlying supposition is: Being alive is better than being dead, right? Not everyone will agree with that. Many friends I've spoken with say that when their time comes, they're willing to go peacefully, without a struggle. I may be one of them, but for now, I'm keeping an open mind.

Whatever the case, there I am, seventy-five years in the future; lying on the table, unconscious but fully repaired, all systems go. My body temperature slowly rises to 98.6 degrees F, my eyes open, and I'm back, just like ol' Khan Noonien Singh, ready to face the backend of the twenty-first century.

The first thing I might do is reread this book to see exactly how accurate it is. It might come in handy as a user's manual for the twenty-second century. The next thing I'd do is check and see if there are any human beings left. I mean if Hans Moravec or Ray Kurzweil are right, they might be few and far between, except for us cryo-regenerates.

Well?

Is any of what I've just described possible? To be blunt . . . no!

Could it be possible? Yes, someday. There are people working on it. Remember I don't create the strange, new worlds, I just explore them. This one's pretty intriguing.

I suppose from the perspective of the present, nearly all futures look impossible. Would any blood-leeching, Dark Age physician have even contemplated heart transplants or brain surgery? Could any citizen of the eighteenth century have looked at a car or a jet or a cell phone and not thought she had lost her mind? Also no. Would Johannes Gutenberg, the man who gave the world books, have been able to make sense of a computer and the World Wide Web, let alone video games and virtual reality? Not likely. But then that's the nature of the future, it is not only stocked with possibilities, it brims with *impossibilities*.

I cannot, therefore, say if these things will come to pass. Nor can I say that we'll have the good sense to get the ethical structures in place to make sure we don't abuse the godlike powers that may accrue in the process. Heaven help us if we decide to use them to turn ourselves into a race of Pamela Andersons and Pierce Brosnans, a planet of homogenized beauties where surprise and diversity have taken a permanent holiday. (No offense intended to the handsome people here singled out.)

Answers to these questions are certainly beyond *my* meager ken. All I can say is that these "could-be"s exist. Unlike *Star*

Trek they are not science fictions. And, as the intellectual pedigree of Ralph Merkle illustrates, we are not dealing with crackpots and fringe snake-oil salesmen. This is serious stuff.

I look at Ralph Merkle across the table over the remains of the meals that we've just obliterated and say, "Well, great. How about if we schedule another lunch, a follow-up to the experiment . . . in, say, a hundred years. Then we can start getting our cholesterol levels back up."

Ralph grins a blinding Merkle grin and says, "You're on. I'll buy."

Toto, I have a feeling we're not in Kansas anymore.

—*Dorothy Gale*

The Wizard of Oz

22

IF IT QUACKS LIKE A DUCK . . .

Virtual reality. Odd juxtaposition of words. Downright oxymoronic when you think about it; like "military intelligence" or "jumbo shrimp." Odd enough, in fact, that I thought I'd look both words up and try to understand what they really mean and where they came from.

It turns out that the word "virtual" has its origins in an old Roman Latin word, *vitus*, meaning strength or virtue. Later *vitus* made its way into Medieval Latin as *virtualis* which means to have certain physical virtues. The words *virtualis* and *Lancelot* probably ended up in a lot of the same medieval sentences. Eventually *virtual* came to mean "seeming to be real," or "having the attributes of reality." So when something is virtual, it seems to mean that all of the evidence points to it being real . . . but you can't actually *prove* that it's real.

Reality itself, on the other hand, is supposed to be that thing that we all experience when we wake up in the morning—the world as we know it: solid, predictable, unmistakably *there!* Reality goes back to *res*, the Roman Latin word for "thing or fact," as in the phrase *in medias res*, which means to start "in the middle of things." Remember that from English class? The story began *in medias res*. We used that technique a lot with *Star Trek;* with the movies more than the television series.

So in the end virtual reality turns out to mean that something or some experience that *isn't* real *seems* real; it *appears* to be unmistakably authentic, when in fact it isn't . . . really.

If this is true, if some experience *seems* so authentic that you can't tell the fake from the genuine article, well, is it still virtual? I mean if it looks like a duck, walks like a duck, quacks like a duck . . . isn't it a duck? Maybe, unless you can't actually pluck it or eat it, or it doesn't *really* bite you when you try.

The Ultimate Playground

Why, Bill, are you yammering on about all of this, you ask? Well, these questions came to mind for two reasons: (1) Virtual reality is a hot topic (and has been for a good ten years) in the worlds of entertainment and science; and (2) *Star Trek* is home to the ultimate virtual reality generator—the holodeck.

The holodeck is without a doubt one of the coolest of *Star Trek*'s imaginary inventions. On the one hand, it's the ultimate playground, a place where you can make anything happen that your little heart desires. Remember when you were a kid and you used to play pretend? Let's play army, let's play cowboys and Indians, let's have a tea party. Well, this is a place where you can *really* play pretend. Want to be Sherlock Holmes, fine. Works for Data. Or Robin Hood? Or a hardboiled detective named Dixon Hill (Captain Picard's favorite). They're all possible. Just load up the holoprogram and you're ready to roll.

The holodeck is also handy for getting work done; a great place for resolving some medical riddle or talking over a nagging problem with someone who knows more about it than you do—

Albert Einstein, for example. Or good for simulating a dangerous situation to see how you perform. And you can do all of this with impunity because the holodeck is programmed to make certain you can never have a lethal accident.

Reading this, you would think that the holodeck actually existed. Of course it doesn't. Nevertheless it sure has turned out to be a brilliant dramatic innovation, like so many of *Star Trek*'s other faux technologies. *The Next Generation*, *Voyager*, and *Deep Space Nine* have all used it very creatively. It not only lets *Star Trek* characters and viewers "believably" enter any number of interesting worlds and merrily pursue the impossible, it reveals a lot about the shows' characters, what they like and what sort of fantasies they prefer. (Okay, I'm guessing that not all of them involve high-minded conversations with Isaac Newton or Leonardo da Vinci, but I'll leave the substance of those fantasies to your own imaginations.)

I'm the first to admit that having something like a holodeck would be mighty handy on a five-year voyage. I wish we had had even an early, hand-crank model on the original series, but, of course, that was out of the question because no one could have conceived of such a thing in those days. I mean, a whole reality generated out of zeroes and ones? The idea of projecting three-dimensional images of people and places and things, in real time, that you could actually interact with? Forget about it. Besides, we didn't have a big enough budget so that we could even fake faking it!

Yet, though we didn't have a holodeck on the old *Constitution*-class *Enterprise*, Gene Roddenberry did have some inkling of it rattling around in his mind even way back in the early days. When I was talking with Bob Justman, the show's first associate producer, and then later its producer, he told me that Roddenberry had always wondered about how a starship crew would get along on such a long mission in space. I mean what do you actually *do* with yourself between battling aliens and solving extraterrestrial mysteries?

Roddenberry knew this had historically been a problem for ships at sea for long voyages. After all, he was heavily influenced by the Horatio Hornblower books and well aware of the historical world-wandering voyages of Charles Darwin and Captain James Cook. So he imagined the day-to-day life of the average *Enterprise* crew member, and it became pretty clear that not every moment on the journey was going to be jam-packed with alien encounters. In fact, he knew that mostly it was just going to be hard, repetitive work. So you have a ship of 430 human beings with normal human emotions and drives and a fair amount of time on their hands. What do they do with themselves when they're not outwitting Klingons or trying to avoid being wiped out by a doomsday machine?

You could see little snippets of concerns like these popping up in the early days of the series. Take the very first pilot—"The Cage," starring Jeffrey Hunter as Captain Christopher Pike. There's an early scene in the show where Pike angrily complains about the stress of commanding long voyages and being constantly away from home. (Suck it up, Pike. This is Starfleet!) Anyhow, this sets up scenes later where Roddenberry dabbles in his own early form of virtual reality, except this variety is not generated electronically. Instead a group of mind-bending Talosians (you remember, those characters with the pulsating craniums) use their telepathic powers to create virtual realities designed to tempt Pike into giving up his command and succumbing to the considerable charms of Vina (played by the lovely Susan Oliver). One of the realities is an idyllic picnic set in the Midwestern countryside with a beaming, all-American Vina at Pike's side. Maybe, telepathed the Talosians, this would sway Pike to give up the *Enterprise*.

Pike, of course, never yields, but think about it. Under different circumstances and with a holodeck available back on the *Enterprise*, this idyllic little scene is probably exactly what Pike would have loaded into the system when he found himself longing for home. He wouldn't have had to decide between home and commanding the *Enterprise*, he could do both.

During the second season of the original series (1967–1968), Gene began seriously toying with an embryonic version of the holodeck. At least I think he was. He envisioned what Stephen Whitfield in *The Making of Star Trek* called "a rather exotic entertainment center." Plans called for future scripts to locate it on the *Enterprise*'s eighth deck, down near the ship's galley. When you read Gene's 1967 description of it, you can sense the origins of the holodeck that would follow twenty years later.

"Man has been too long a part of Earth to be too long separated," he wrote. "Therefore we intend to build a simulated 'outdoor' recreation area which gives a realistic feeling of sky, breezes, plants, fountains, and so forth."

Obviously going "back home" was something that Gene felt crew members would want to do, even if they could only do it by playing an elaborate game of "pretend." Okay, I know, it's probably true that Vulcans would only need to return home during *Pon farr*, and you can hazard your own guesses as to what Cardassians or Klingons or Bajorans might long for, but the point was that Starfleet personnel were human (or at least human*oid*); they had emotional needs that had to be fulfilled and if they weren't they would lose their minds, or worse, stage a mutiny. Either way it would have made life damned uncomfortable for the likes of me.

But as I say, we were "without holodeck" in the original series, despite Gene's plans for a season-three playground. By the time that final season rolled around, *Star Trek* budgets were tight. Big special-effects-laden entertainment centers were certainly not, as it were, in the stars.

The Next Generation

So now fast-forward two decades. Obviously between the last episode of the original series and the first episode of *The Next Generation*, the "playground" issue had continued to nag Gene. During early story conferences for *The Next Generation*, the subject came up again, except this time it was Bob Justman who raised the issue. Bob was back on the job as producer, and he knew the question of how crew members would pass their spare time was going to come up. His suggestion was that the *Enterprise* be outfitted with what he called "video walls." Maybe, he suggested, this would help to creatively supply some of the diversions that crew members would need as they bounced among the stars. Want a walk along the beach? Zap, the video wall's a beach. Want to go to the wine country of France? Fine. You're in France. The whole idea was that you would provide crew members with a kind of omnimax effect, a convincing, if artificial, sense of place. It wasn't yet the interactive world of the holodeck, but it was a beginning.

According to Bob and Michael Okuda, another early member of *The Next Generation* team (also author with his wife Denise of *The Star Trek Encyclopedia*), the video wall concept was massaged over the course of several meetings until, inch by inch, the holodeck emerged. Crew members would be able to enter a self-created digital world that not only looked real, but also acted real.

This was a *very, very* cool idea for its time. In fact, the holodeck was far enough ahead of the curve that it actually predated the phrase "virtual reality," which didn't come into general use until the late 1980s. Not that scientists hadn't been grappling with the concept. They had. In fact, according to Jaron Lanier, the man credited with coining the phrase, Ivan Sutherland succeeded in creating the first "virtual world" way back in 1968 (just a year after our last season). Sutherland is generally considered the father of computer graphics, says Lanier, and he managed to create his virtual world by engineering a bulky head-mounted display that had a couple of small screens positioned in front of your eyes. It wasn't really all that real and it took the most powerful computers in the world to generate the displays, but it was a beginning.

Later researchers at NASA Ames started to experiment with ways to virtually place you in another, *actual* environment, not a pretend one. They called this "telepresence" and experiments included putting you on the moon or the plains of Mars, digitally. The idea was that you would be linked to a robot in one of those distant locations and therefore be "present" there, without actually having to *go* there. Again, a step in the right direction, but a long way from the holodeck. It wasn't until the late 1980s that telepresence expanded to include the creation of whole environments that were *entirely* digital; artificial places completely fashioned out of nothing more than numbers.

Going all the way back to those early *The Next Generation* brainstorming and production meetings, the holodeck pushed the envelope on digital environments (and it's grown more ambitious over the years, getting upgrades from series to series and movie to movie). A holodeck is a place where digits are woven together so powerfully that it is virtually (sorry) impossible to tell what is real and what isn't. It's science, or more precisely science fiction, at its most magical. Load a program into the holodeck and you can talk with digital re-creations of people (living, imagined, historical) as if they were as alive and as *there* as you are. You can pick up objects, bump into walls, hop on a horse and gallop into the sunset, all without ever leaving the confines of the suite. Yet everything you experience will seem perfectly real. This is powerful. So powerful that someone once remarked that after a holodeck is actually invented, no one would ever work again . . . unless it was to make more money to pay for new holodeck programs.

But we're not in any danger of that happening, not yet, anyhow. After all, how would you actually pull something this spectacularly, artificially realistic off? Of course, I haven't a clue, but I knew there were sources—my copy of *Star Trek: The Next Generation Technical Manual*, for example. So I pulled it out to see what it had to say.

I always marvel at how full-blooded the *Star Trek* universe has become over the last thirty-five years. It really is unprecedented! No imaginary world is as well documented or as thoroughly scrutinized as *Star Trek*'s. I still sometimes have a hard time comprehending the whole fabulous phenomenon. Its imaginary stories span more than three centuries (not counting the time-travel segments), several universes, hundreds of characters, scores of worlds, and whole flocks of technologies, every last one of them the figments of very creative imaginations. Each of these has been thoroughly described, discussed, and dissected, none of them more than the technologies.

Maybe that's why when I read the *Technical Manual*'s description of holodeck technology, I found myself grinning from ear to ear. It *sounded* so real, yet, of course, it can't be, otherwise we would already be swimming in holodecks. You wouldn't be *reading* this book, instead you and I would be finishing up a nice dinner at your favorite restaurant and enjoying a fabulous Frangelico tiramisu dessert as we continued our conversation deep into the digital night. I, of course, wouldn't actually be there. I'd simply be a fabrication of the holodeck, but I would sure as hell *seem* to be there.

And if you asked me, "Bill, how could something like this be possible?" I'd pull out my trusty copy of *Star Trek: The Next*

Generation Technical Manual and read you the following passage on pages 156 and 157:

> The holodeck utilizes two main subsystems, the holographic imagery subsystem and the matter conversion subsystem. The holographic imagery subsection creates the realistic background environments. The matter conversion subsystem creates physical "props" from the starship's central raw matter supplies. Under normal conditions, a participant in a holodeck simulation should not be able to detect differences between a real object and a simulated one. . . .
>
> The basic mechanism behind the holodeck is the omnidirectional holodiode (OHD). The OHD comprises two types of microminiature devices that project a variety of special forcefields. The density of the OHDs is 400 per cm^2, only slightly less than the active visual matrix of a multilayer display panel, and powered by standard medium-duty electro plasma taps. Entire walls are covered with OHDs, manufactured in an inexpensive wide-roll circuit printing process. . . . The primary subprocessor/emitter materials include keiyurium, silicon animide, and superconducting DiBe<2>Cu 732.

Sounds damn good, doesn't it?

But how close are we, really?

Creating the Virtually Real

Decades off, says Peter Norvig, chief of computational sciences at NASA Ames in California. When pushed to hazard a guess, his blue eyes gaze for a while at something beyond the wall of his office and then he says, "Thirty or forty years. It's the creation of digital objects that's tricky. How do you give something that's a digital fabrication weight and body and heft?"

Giving "objects" that are really nothing more than complex representations of zeros and ones the feeling that they are actually there requires what is known as "haptic feedback."

When the scientists and researchers at NASA Ames Research Center first threw this term at me, naturally I was clueless. No matter how much I wrinkled my brow, I couldn't even hazard a guess. I just stopped in mid-nod, looking befuddled, and said, "Haptic?"

Haptic, I was patiently informed, is all about the sense of touch. We take for granted the complexity and importance of touch, I was told. When you pick up a glass, you grip it hard enough that it won't slip from your hands, but not so hard that you crush it, mangling your digits in the process. Our hands, feet, our whole bodies are sensitive to the feedback that real objects give us so that we handle them in a way that gets the results we want, no more and no less. You apply one kind of pressure if you're trying to roll a boulder, another if you want to hug a loved one. Confusing one with the other could be dangerous.

The whole question of "hapticness" is in the most rudimentary possible stages right now when it comes to virtual reality, but at NASA they're working on it. I got a taste of it when I was flying that virtual F-15 jet fighter. The joystick, for example, pushed against my hand if I tried to move the jet to the right when it wanted to barrel left. I got another taste of "force field" technology when I was allowed to get into a big harness and seat, and virtually dock an imaginary spaceshuttle with a make-believe International Space Station. I could sense the movement of the ship as I pushed various thrusters, and when I finally managed to successfully dock, I could "feel" the marriage of the two nonexistent crafts through the controls in my hands as they locked together.

NASA is even creating a system designed to help doctors operate in extremely small and delicate areas of the body like the brain where the tiniest miscalculation could be lethal. I was allowed to try this out too. These systems are a variation on the concept of telepresence, and what they do is make it possible for surgeons to experience the body as if they were themselves not much taller than a thousandth of an inch. It's as if they were characters in that old science fiction movie *Fantastic Voyage*. You know, the one where a ship is shrunk to the size of a bacterium and injected into a human body so they can perform brain surgery on the patient.

Anyhow, soon doctors using these "virtual" surgical techniques will be able to make their incisions by operating a digital scalpel which will in turn be operated by a computer controlled robot. The virtual reality that the surgeon enters will then allow her to "see" what the knife is doing in such detail she'll be able to notice the difference between each layer of skin! It will also provide far more sensitive haptic feedback far sooner than her own hand would. She'll know precisely where an important blood vessel or bone is, and exactly where the good tissue ends and the damaged tissue begins.

But again, as amazing as these experiences at NASA were, I was looking at a monitor; I wasn't really being submerged in a completely digital world where the real and the unreal were indistinguishable. And that's what I wanted, something really (or is it virtually) holodeck-like. So once again, ever focused on our perilous mission, Chip and I took our old-fashioned, molecular selves to Carnegie Mellon University's Entertainment Technology Center where we planned to get totally and digitally doused.

Reality is that which, when you stop believing in it,
does not go away.

—Philip K. Dick

Human kind
Cannot bear very much reality.

—T. S. Eliot

23

VIRTUALLY POSSIBLE

If you're planning to get immersed, digitally, Randy Pausch is a good man to see. He's a lanky, dark-haired scientist who, despite his forty years, could pass for one of his own graduate students. (Perhaps this is a clever digital special effect?) With drama professor Don Marinelli, he co-directs Carnegie Mellon's Entertainment Technology Center.

When we walked into the class's virtual reality lab, my first impression was that it looked like someone tossed an eight-year-old's bedroom, an amusement park arcade, and geek heaven into a blender. The walls were covered with stuffed animals, a menagerie of grinning, wide-eyed lions and tigers and bears. "We won them at the amusement park," explained one grad student. The ceiling hung with wires and diodes and sensors, the names of which I could never possibly pronounce let alone comprehend. These, I later found out, were designed to track and sense the location and position of yours truly once he had entered the various virtual realities soon to be made available.

Nearly every section of the room sported clusters of humming computers and monitors. Nothing terribly out of this

The bridge of the *Enterprise* the way I remember it from 1966.

world, just the sorts of boxes you would find in any run-of-the-mill gadget store. But there were a lot of them. "Everything we do we do with a standard thousand-dollar home computer," says Randy. "We're working on creating software tools that will allow anyone to create their own virtual realities."

Now *there's* a thought.

Back on the Bridge

Randy marches toward the back of the room and stands on a big X marked on the carpet with duct tape.

"Come stand over here," he says with a grin. "This is our first demo. I think you might find this environment familiar."

I look at him. Why do I always feel like I'm about to be the butt of a practical joke in situations like this?

"What's that?" I ask, pointing at the thing in his hand.

He looks at it and then back at me as if to say, How could you *not* know what this is?

"This is a head-mounted display. There are screens inside,

very small screens that project the images of the virtual world you are about to enter. It gives you the feeling that you are actually in the place that we have created. But first you have to stand on the X."

So I walk over and put on the head-mounted display. To the other fifteen bemused students in the room, I must look like some bizarre insect with this thing on my head, but I can't see me. What I see is the world that has been created by the students and their computers. And I have to admit, it looks pretty familiar. It's the original bridge of the starship *Enterprise!*

"Well, I recognize this," I say, grinning. "But where are Uhura, Spock, and Sulu?"

"Well, this is just a simple demonstration," says Randy. "Something we threw together over the weekend. Creating virtual places is actually relatively easy. Creating virtual people, that's something else."

The bridge of the *Enterprise* the way I experienced it at Carnegie Mellon Entertainment Technology Center. It's not the "real" thing, but it seemed real enough to me when I made my virtual visit. *(Carnegie Mellon Entertainment Technology Lab)*

So I look around my pretend world. There behind me is my trusty captain's chair where I had planted my tail for three years and commanded so many missions.

"And there behind my chair," I said with some surprise, "are the turbolift doors that I smacked my nose into every time the grip who operated them didn't pull the rope in time. Hurts just to *think* about it." (General laughter from the invisible students.)

As virtual realities go, this was more real than most, at least most of the others *I* had tried. But it wasn't because of the fidelity of the images, which were far from smooth and realistic looking. It was because I felt as though I really was in a *place*; totally immersed. When I turned my head, the bridge swept before me with all of its consoles and lights. When I looked down I saw the floor and the chair. When I turned, there was the viewscreen where so many imagined planets and aliens and enemy ships had once appeared. In fact, the general impression was so real that when the *Enterprise*'s red alert horn started blaring, I nearly jumped out of my skin! For just a split second I was sure we were actually under attack! But then I realized we couldn't be because this wasn't the *real* set, let alone a *real* starship. Not only that, this was a virtual version of an imaginary ship that wasn't even supposed to exist for another three hundred years! So I got a grip and managed to stifle the urge to jump into my digitally manufactured captain's chair and bark, "Mr. Sulu! Evasive maneuvers!"

I had been had.

I took the helmet off and looked at all of the chuckling graduate students who suddenly appeared before me. "You got me," I said, grinning. "Lucky for you it was only a *virtual* emergency. I could have started firing my phaser."

"Well, we're just in the parlor trick stage of virtual reality," said Randy. "You know, just a few cutesy little demos to give you the sensation that things are really happening."

"Well, it worked!"

It worked, Randy explained, because the software and headset combined to create a real sense of motion and place in the virtual world that was consistent with my own movements. Because the two were in sync my brain was fooled into thinking that I was actually moving around in a space that didn't even exist. If I walked, some things moved closer, some receded. If I turned my head, the scene changed in real time. It didn't matter that the images were cartoony, the experience was accurate and that was good enough to provide more than a taste of reality.

"Ready for another demo?" asked Randy.

How could I resist?

This time I planted the headset on my noggin and slipped on a special glove loaded with sensors designed to provide me with a virtual hand in the virtual world. With the headset on I found myself standing in the center of a make-believe courtyard surrounded by buildings.

Randy then asked me to step up on a platform. Did I see it? Yes, I said looking down. There one sat, a cartoon version of a platform which represented an actual one that existed in the real world. So when I stepped up onto the digital version, I felt solid wood beneath my feet. Very weird.

"We're going to try a little experiment," I heard Randy say. And slowly the platform, the virtual one, begins to rise ten feet or so. Of course to everyone looking at me, I am just this guy standing on a box in a room full of stuffed animals, but to me, I was perching precariously one story above a digital courtyard. Then magically a light bulb appears out of nowhere just within arms' reach.

"Does that mean I just had a great idea?"

Disembodied laughter.

"Can you grab the light bulb?" asks Randy.

Like I would know.

But I reach up anyhow, and low and behold the virtual hand (strangely unattached to an arm) moves toward the virtual bulb and grasps it. I pull it over.

"Now," says Randy's voice coming from nowhere (now I know what schizophrenics feel like), "drop it."

Okay, I figure. It's only pretend anyhow. So I open my hand and the imaginary bulb drops like a stone to the digital cement below, making a pop-tinkling sound the moment it hits; just what you'd expect from a light bulb. The volume of the sound of it hitting the pavement was even just right, as if it were actually about ten feet away from me.

"Now we're going to go a little higher," says the voice.

Now I find myself rising higher into the artificial sky. The buildings in the courtyard recede. The pavement below grows more distant, and then all of the primal reflexes start kicking in. I'm standing fifty feet above the ground, and my brain is screaming, "Danger! Danger!"

I mean I know this isn't real, but as I teeter at the edge of the platform, a primal fear of falling grips me. It's hard to shake the feeling that I am really perched five stories above the ground. Now a new light bulb appears. Another great idea, like, Get me out of here! The voice asks me to grab the bulb again and let it drop. So I do, with every autonomic reflex, every hormone screaming to be careful not to slip off the platform and end up a virtual grease spot on the pavement below. The little bulb plummets, seeming to take forever to hit the ground. When it does, I hear a very distant popping sound.

Now my mind is bifurcating. On the one hand, all of the primal drives, all of the physical cues being fed to my brain by the computer tell me that I am standing on a tiny little platform fifty feet above rock solid pavement. On the other hand, the logical part of my mind tells me that I am simply standing on a box no more than six inches high and if I step off I'll be on solid ground, safe and sound.

As if reading my mind, Randy says, "Now step off the box."

The primal and logical parts of my brain debated the suggestion. And you know what? I did step off . . . but not until I removed the headset.

I may be adventurous, but I'm not stupid.

Round of applause. I bow my best stage bow, and then look up at all of the students.

"Okay, how do you do all of this?"

"It's simple, really," Pausch says.

"Well, simple to you maybe . . ."

Tricks of the VR Trade

It turns out that Randy Pausch's "parlor game" virtual reality feels as immersive as it does because it completely fools the brain on the most primal level. First, of course, there is the headset which places you "in" the world, rather than making you a spectator. The headset envelops your whole field of vision and that eliminates any peripheral visual distractions that the real world would provide *if* you could see it. As far as your brain is concerned, at least the visual, primal part of it, you're not where logic tells you you should be, you're somewhere else. It's a very interesting feeling. Then there is the tracking device which ensures that as you move through the world, it changes in a way that is consistent with what your senses expect. This completes the trickery because when you turn your head left, or step to your right, the tracker makes sure the computer projects the image you should see just as your brain and eyes would in the real world. See, it's just a very effective shill game.

"The tracker is nothing more than a fancy compass," says Randy. "It 'talks' with the headset and describes which way you are turning your head and where you are located. It continually transmits that information to the computer, which in turn generates the images for the virtual world."

Okay, I get that. It tells the virtual world that I'm there, and it also tells the virtual world which way I am moving my head, what I'm looking at, where I am walking. In short, it senses my real self so it can then place that self accurately in the computer-generated world.

For example, explains Randy, when I had the glove on, the tracker was calculating the location of my hand because the glove had transformed it into a digital object. That's how I was able to grab a light bulb that was, in reality, no more than an image projected by the little television cameras inside the headset. The tracker and the glove connected the analog world to the digital one the moment I put the headset on.

But how do you create the light bulb, not to mention the rest of the objects that populate this dream world? Turns out it's a little bit like creating a digital special effect, except in this case you can interact with the effect in three dimensions.

"First," Randy explains as he walks me over to a group of computers and students, "the artists and architects on the team create the virtual 'structures' in the world using software that generates objects in three dimensions, or 3-D. That means they create a three-dimensional mathematical description of each piece of a scene—a chair, a building, a console. They look like objects, but of course they're just images. Artists first draw a wire frame. The wire frame is a representation of the underlying structure of the objects, kind of like a skeleton. The frame is literally an illustration of a mathematical description of the object."

But what is amazing about this, at least to me, is that these aren't just simple two-dimensional renderings that look like three-dimensional objects. When an artist uses this software to "draw" something, it apparently also stores all of the mathematical information that the computer needs to redraw that object from any imaginable angle.

"Doesn't matter whether you are standing on it, lying under it, walking by it, approaching it, or leaving it behind," says Randy. "The mathematical description provides all of the raw material that the computer needs to manufacture a digital version of, say, the captain's chair on the bridge of the *Enterprise* that looks the way it should based on where you are located in relation to it.

"What the computer is doing unbelievably fast is getting the information from the tracker and feeding your movements and location into the model and then instantly redrawing all of the objects in the scene to correspond with what you're doing. It does this in one thirtieth of a second, and because it does it in one thirtieth of a second, it *feels* real because that's how fast your eyes need for it to happen to deliver the illusion of reality."

All of this works together to create the mirage, the "virtuality," of the whole experience. This is crucial because you can't create an illusion of reality if the virtual environment remains unchanged as you move through it. In fact if it remained unchanged you *wouldn't* be moving through it. It would be as though you were walking around a painting in a museum. No static image, no matter how arresting it is, can ever leap from its two-dimensional space and morph as you move.

Can We Get Real?

Okay, fine, now I understand what the head-mounted visual display and the tracker do; I comprehend that feedback is important. But what about creating environments that not only act real, but look really, well, real?

Not in the cards right now, says Randy. "The kind of raw

computing power needed to conjure a digital reality that is so full-blooded that it is indistinguishable from the true and authentic thing is not possible, not today, not yet. Even when it comes to visual images, we're a ways off from the holodeck."

Why? Well, here are a few of the obstacles. Our eyes can resolve objects down to fractions of an inch. That means for virtual objects to look real, they have to be extremely detailed, at least as detailed as what you see in an animated movie like *Toy Story*, or an extremely realistic special effect.

Okay, fine, you say, make the objects in the virtual world look like great special effects. No can do, says Randy, because when you add all of that detail, you are also adding immense amounts of information to each and every virtual object. That's bad enough, but in a virtual world you also have to redraw all of those objects in the same detail, *and* you have to do it within one thirtieth of a second—the speed of reality. This, explains Randy, takes some industrial-strength processing power. Even the most muscle-bound computers around can't redraw high resolution objects fast enough to keep up with the simplest human movements and gestures, not right now.

That's problem number one.

Going Digital

"Then there is the issue of how you actually place your entire self into the virtual world," says Randy, sitting on a big desk full of computers. Easy, you say, open the holodeck doors and walk in.

Someday, maybe, but not today. The big problem here is that you are a molecular object, not a digital one, whereas the world that you are entering is one hundred percent digits. "Virtual worlds are nothing more than complex representations of numbers that happen to look like . . . well, whatever you want to make them look like," says Randy.

That means that somehow *you* also have to end up as a mathematical . . . thing. In other words, you have to go digital.

In my virtual escapade at Carnegie Mellon, this problem was partially solved by my wearing the head-mount, which (apart from making me look goofy and my bad hair day worse) partially digitized my head. The glove, when I put that on, did the same thing for my hand. But what about the rest of me?

Compare this with *Star Trek*'s holodeck and you find just how differently it and the kind of virtual reality that folks like Randy Pausch are tinkering with is. In the holodeck you don't have to become digital because everything around you is a projected, three-dimensional hologram. Instead of you entering the digital world, the digital world is projected into the molecular one. That's why you never see anyone in *Star Trek* putting on a speck of special equipment before they enter the holodeck. Projecting virtual reality right into the real world, however, is unbelievably difficult, unless, of course, you're just pretending, which is what *Star Trek* is really good at.

But just for kicks, how would you do it?

Holographs have been around for decades. But most of what you've seen represent a single image. It *appears* to exist in three dimensions, but it is still static and unmoving. Even sophisticated holograms can't walk toward you like a holodeck-generated Isaac Newton or *Voyager*'s Doctor, because, as the experts have told me, this is just asking more than any computer can do right now. When Chip and I visited the MIT Media Lab we did see a very brief and very small projection of a three-dimensional moving image, but it was a long way from perfectly realistic (folks at the Media Lab would be the first to agree). And for now that means that the digitize-yourself-approach seems to be the path most scientists are going to take toward the holodeck.

So, you're wondering, can you do it? Well, maybe, if you're willing to slip into a digital body suit which is something like a cloth version of a wet suit, except its plastered with sensors that place not simply your eyes and head and an occasional limb in the virtual world, but every part of you.

At the University of North Carolina at Chapel Hill, digital body suits are just what scientists are using to explore the strange new worlds of virtual reality. In fact they have built a twenty-six by eighteen foot room that is laced with what they call "a grid of infrared-light-emitting diodes," every one of them mounted on the ceiling. Wherever you walk, however you move, whichever way you turn your gaze, the scene in your head-mounted display changes accordingly at the stupefying rate of 1500 times a second as the diodes and sensors and computers all chatter away and update the digital world.

Seems to me that something like this could get real interesting, real fast. Put a couple of people in the room, appropriately outfitted, and they could do any number of amazing things. Play tennis without a tennis court; explore an imaginary virtual world as wild as Alice's Wonderland; pretend they are two crew members of the *Enterprise* and visit Sigma Iotia II. It's not the holodeck, but it's a start. (And I like the dimensions of the room, which I suspect aren't all that different from the size of a holodeck.)

Virtually Human?

There's another third, huge issue that VR researchers struggle with. How to create other virtual characters, real full-blooded "people" who happen not to be actually real, or are they? This is, after all, one of the things that makes the holodeck the holodeck, right? It generates realistic, digital versions of other people and creatures. You can interact with them, talk, dance, you name it and they interact right back in ways that make sense.

That, of course, is the truly tricky part. It's one thing to project an image that looks real, it's another to get them to *act* real, like, say, the Doctor in *Voyager* or the characters in Captain Picard's favorite holofantasy, the world of Dixon Hill. However, as you know from reading the AI section of this book, creating characters that can act human is not easy. In fact, says Randy, if you could create characters—historical, imaginary, or personal—that acted absolutely realistically, they would all have passed the Turing Test. And passing the Turing Test means that you would have created an artificially intelligent creature with emotions and instincts and memories that *seem* completely human; in effect you would have created another sentient being.

"Well, that's a whole other problem," says Randy.

However, what we might be able to do while we're trying to figure out how to create virtual humans is create pretty interesting virtual worlds where we can hold business meetings or family reunions, or play sophisticated role-playing games with other digitally outfitted human beings. You know, "Okay, honey, tonight I'll play Captain Kirk and you play *all* of Mudd's women. Just flick this button here when you need to change identities." Imagine the possibilities! And they think the Internet is cutting into work time.

Solid Problem

Finally, the last thorny VR issue: How, in a universe that is fashioned out of something as ephemeral as pure information, do you create solid objects? This is like making a car that you can get in and drive around, out of the words that *describe* a car.

Once again it appears the holodeck has an advantage over science. With the holodeck, for example, you can just bring your own objects in, something you can't do in a Randy Pausch–style virtual world, unless it's wearing a digital body suit.

Actually when entering the holodeck, the only object you really have to bring inside is yourself because it supplies all of the solid objects you need. How? How can you drink a virtual drink or help yourself to a fake canapé, or how do you create arrows that can be notched and actually fly through the air?

The creators of *Star Trek: The Next Generation* addressed this problem in a couple of ingenious ways. Remember that the walls of the holodeck are lined with hundreds of millions of those little OHDs, omnidirectional holodiodes? Together these holodiodes not only beam the holographs into the holodeck, they create the impression that certain objects have weight and mass and texture by emitting a small force field. According to the *Star Trek: The Next Generation Technical Manual*, these tiny force fields unite in all of the appropriate places, right in thin air, to give the impression that a real rock or gun or chair is actually there.

Could we do this, really? No, not today, maybe not ever. It requires mastering antigravity, but I suppose if you're riding a ship that can warp the fabric of space and time, what's to stop you from creating virtual objects with millions of tiny force fields? Small problem.

It turns out there is another handy way to create objects in the holodeck. Use an onboard version of the replicator and beam them in, just like Scotty used to beam us all from place to place with the transporter. Except instead of beaming one group of molecules from one place and rearranging them in the same exact form somewhere else, the holodeck gathers up a passel of molecules out of the air and then arranges them in such a way that they become whatever you want them to become, say the '57 Chevy in the "Lifesigns" episode of *Voyager*. (My personal choice? A 2001 Harley-Davidson Electra Glide Fantasy, but whatever starts your engine.)

I suppose the thinking behind this approach is: Why simulate when you can replicate. The only limit is your imagination . . . and the number of freely available molecules hanging around.

But again *Star Trek*'s solutions to scientific problems aren't always the same as those of science. After all, *Star Trek* has created a kind of parallel universe over the years populated by its own imaginary technologies. When it comes time to create the next generation of mythical technologies, it makes sense to build on what is already there. In fact, there's no choice; otherwise *Star Trek*'s own world would lose its integrity and consistency, and then where would we be? Civilization as we know it would come to a screeching halt.

On the other hand when it comes to creating real virtual worlds (is that possible?), then the door is wide open, and scientists can consider solutions that even the creators of *Star Trek* can't. Foglets, for example.

Foglets could solve the solid-objects-in-a-digital-world riddle, and even if they can't they're just too cool not to mention. There is one small problem. In order to create Foglets you have to have the great, yet-to-be-invented "ology" of the future: nanotechnology.

Foglets come by way of J. Storrs Hall, a computer scientist who left Rutgers to work at, you guessed it, the Institute for Molecular Manufacturing, which is itself affiliated with Eric Drexler's Foresight Institute. (The nanocommunity is apparently pretty small. But that makes sense.)

Storrs first wrote about what he calls Foglets, or Utility Fog, in *Extropy* magazine a couple of years back.* He describes them

---

**Extropy* is a publication that walks the thin line separating advanced scientific theory and science fiction. While the magazine definitely pushes the envelope of acceptable science, it's a long way from wild fantasy. Everything in it is scientifically based, and its contributors include an enviable list of talented researchers and thinkers. It's just not *Physical Review* or even *Scientific American*. It *is* a place where scientists can limber up some of their more creative ideas, even as they work to nudge them into the realms of actual applied research.

"The Utility Fog is a very simple extension of nanotechnology," explains Hall. "Suppose, instead of building the object you want atom by atom, the tiny robots linked their arms together to form a solid mass in the shape of the object you wanted? Then, when you got tired of your avant-garde coffee table, the robots could simply shift around a little and provide you with an elegant Queen Anne piece instead." *(J. Storrs Hall)*

as invisible robots about ten microns across, which is to say about the size of a human cell. By nanotechnological standards this is gargantuan, ten thousand times the size of a nanomachine, but, says Hall, that's okay because you need nanomachines to build the Foglets in the first place. And once you have built a few bazillion Foglets you get Utility Fog. And the cool thing about Utility Fog is that it will also pack an enormous amount of computing power because each Foglet is a very sophisticated robot. For another, Utility Fog is networked so every Foglet can talk to every other Foglet.

All of this makes these invisible gadgets particularly good at blurring the lines between the molecular and digital worlds, and that is precisely what we want when exploring how we might create a holodecklike technology. Since each Foglet is a computer, it is digital; but since each is also made of matter, it is also molecular.

To manage this trick, says Hall, Foglets have a very specific design. At their centers is a dodecahedron, a kind of ball with twelve flat surfaces. From each of these surfaces protrudes a mechanical arm, nothing fancy, just a straight appendage with grippers attached at the end. These grippers surround a socket not unlike the socket into which you clip your telephone jack at home. That's called a communications port. All of these grippers ensure that each Foglet can connect to another Foglet in nearly unlimited ways, and, the way I see it, that's where the connection to the holodeck lies.

Hall didn't have the holodeck in mind when he conceived this little device, but I suspect Starfleet would love to get their hands on it nevertheless. Imagine you walk onto a holodeck. The chamber is loaded with invisible, floating Foglets; billions of them stealthily riding the currents of air more easily than the smallest mote of dust. Except unlike dust they are all chattering away, keeping track, wirelessly reaching out and touching one another.

Now let's assume that your particular holodeck fantasy involves playing a grand piano at the Vienna Opera House in 1812. As soon as the holodeck boots up, billions of Foglets instantly know what to do. They flow together, lock their little robotic arms and, right before your eyes, form something that looks very much like a grand piano. Now in truth this thing isn't actually a grand piano. In fact if you played it, it wouldn't really have the acoustic integrity of a Yamaha or a Baldwin or, for that matter, a broken down honky-tonk upright. That's because Foglet-formed objects are only about one-tenth the density of real objects, says Hall. But again they are objects that consist of billions of extremely powerful computers so this particular grand piano, fraudulent as it actually is, would be a machine capable of digitally *faking* the sound of a real baby grand. It wouldn't operate like a vintage piano, but the result would be the same (assuming you know how to play it).

"Utility Fog pretty much makes it possible to create any environment you want just by having the robots link up in whatever shape you like," says Hall. "Make all of the walls in your house out of Fog and you can change the floor plan anytime you

Each Foglet has twelve arms. The arms telescope rather than having joints, and they swivel on a universal joint at their base. Any two Foglets can "clasp hands," which forms a relatively rigid connection. Their size (about ten times the size of a human cell) and design allow them to form whatever shape you can imagine. Could Foglets someday deliver holodeck-like objects that seem as real as the real thing? *(J. Storrs Hall)*

like, any way you like. Every day you get a new house. Make a Fog floor and it'll never get dirty. It'll look like hardwood but feel like foam rubber, and grow any kind of furniture you can imagine. All you have to do is tell the little robots what you want; except for food, they can't help you there." (So much for Ktarian eggs.)

From where I sit—and this is a long way off from a Ph.D. in computer science—it looks to me like Utility Fog, or something very similar, would pretty much solve the holodeck's "solid object" problem. It doesn't require force fields or molecular transporters. Instead it just locks together like LEGOS and forms whatever you need (as long as you don't need a Big Mac).

Impossible? Maybe, but we are already building machines that are micron sized. So who knows? Foglets could even bring a certain brand of virtual reality right into your personal life. Imagine having a few billion hanging around your house. Life would become a Disney animated feature. You might have a few hundred thousand Foglets form up into a mechanical hand, for example, grab a beer out of the fridge, and bring it to you on the Foglet-formed couch. Then you might command a battalion of them to go sweep up the floor . . . after another group of Foglets formed up into a broom. Or better yet, while you sleep program them to scour every nook and cranny in the house, micron by micron, and toss the debris in the garbage.

The point is this is a very slick concept. In fact if Hall could ever actually design and build his Foglets, I'm sure Starfleet would happily pay him a hefty licensing fee. I mean, why not? It would save them a barrel of time and energy, not to mention whatever kind of money they spend in the twenty-third century.

Of course I wouldn't mind getting a few stock options in the company first. I mean this could be big! Really . . . I mean, *virtually* big!

But all shall be well, all shall be well and all manner of things shall be well.

—Julian of Norwich

EPILOGUE

WHAT'S A FUTURE FOR?

It's been a helluva ride. I don't think the *Enterprise* herself could have delivered much better. We sat down at the table of "around-the-bend" and got fed a smorgasbord.

Personally, I'm stuffed.

We started this book in one millennium and finished it in another. That's fitting I think; a metaphor for just how fast the world is changing; how rapidly todays morph into tomorrows.

You don't have to read *Scientific American* to see how quickly change is coming at us (especially if you actually read this book). Just pick up a recent issue of *Time* magazine or your local paper. Choose any headline at random and chances are it will tell you how some new innovation is about to upend the present and send us careening off to parts unknown. Every few years now we double the sum total of human knowledge. And each year that "doubling time" grows shorter. I read somewhere that *The New York Times* contains more information today in one edition than the average seventeenth-century scholar digested in a lifetime. It's the Tribble Effect at work.

So what do we do when the future is moving our way so fast

that we're hard-pressed to distinguish it from the present? Personally it makes me wonder what a future is for in the first place. I've come to this conclusion: Futures are for dreaming. We all hope that the life we are living today will somehow be better tomorrow, or at least no worse. Thanks to the mind-bending progress we seem to be making we have the luxury of hoping that tomorrow's solutions will eradicate today's crises.

Dreams like these weren't always possible. The future, at least the way we see it today, is a relatively new concept. Just a few short centuries ago nothing much changed from day to day or, for that matter, from generation to generation. Yesterday looked pretty much like today, and tomorrow, as a result, did as well. There must have been a severe shortage of hope in a world like that. But there must have also been a certain sense of security.

But whatever security the status quo once offered is all gone now because with all of the change we're seeing—and all of the hope that goes with it—we are also getting a fist full of speed and complexity. We're like the driver behind the wheel of a car that's suddenly accelerated from zero to 150 miles an hour in the space of a few seconds. Not only that but we're not sure how to operate the damn thing! How do we keep it on the road, make the right turns and the best decisions when we're just learning to drive? Any starship captain can tell that the course for the future is always set in the present, but at this speed how do we know how to calculate the correct trajectories?

Where's Sulu when you need him?

Throughout the course of this adventure, with each new innovation I've learned about, I've wondered about our ability to proceed. How has the human race managed to get itself tangled in so much complexity? The answer, of course, is the Wonderment Disease, our genetic affliction. We keep looking over the next horizon and shoving the frontier forward. We now find ourselves staring in the face of multiple frontiers. What happens if we create highly intelligent machines and robots? What happens when we are all enveloped in a bitstorm that makes us just one more creature in the information ecology? How do we handle cloning, or the possibility of life with no apparent end, or forays into parallel universes fashioned out of the transcendent tapestries of zeros and ones? These things are no longer imaginary, they are knocking at the door. Now what?

With our heads spinning, our first instinct might be to attack the technology. (I certainly have.) Sometimes I wonder if we invent the technology or the technology invents us. It's both, I think. Look at how deeply each breakthrough reshapes us, and then how, in turn, we take each breakthrough to make another. So the real question becomes how will we put the seismic innovations we excel at to work? Will we use them to amplify our humanity, or blunt it? To empower us or confuse us? To fulfill dreams or create nightmares? Can we drive at high speed?

When I think about these things, I find myself looking back over my shoulder at *Star Trek,* and I have to smile because it's in *Star Trek* that I see hope. One of the great and enduring appeals of *Star Trek* is its optimism and its humanity. It painted a portrait of a world, an entire galaxy, where, instead of grunting and shuffling through the debris of postapocalyptic misery we've learned to forge a workable alliance between technology and the best part of ourselves to build a better future.

I suspect that one of the purposes of science fiction is to let us play out our nightmares and our dreams in the theater of the future *before* we turn them into reality. It's an imaginary place where we can explore the mistakes we might make *before* we make them, a place where we can try new possibilities on for size. Maybe in that way science fiction both inspires us and warns us: The future can be better, but careful what you create; careful what fire you play with and what you let out of Pandora's box, because there's no going back.

I'd like to think we will create the kind of future that *Star Trek* imagined. I'd like to think we will fulfill dreams rather than erect nightmares. I'm not saying that the world in which we someday live will be like *Star Trek* in every detail. It won't be. But I do hope it will be humane and principled. That's what's at *Star Trek*'s foundation: a bedrock faith in humans to do, ultimately, the right thing.

The people we talked to for this book want that kind of world too, I think, partly because they have also been influenced by *Star Trek*.

The whole process has come full circle.

But the future isn't really up to them. Ultimately it's up to all of us to choose which directions we take. *We*, not science with a capital "S," will have to plot the course. There won't be any James Kirk to make the decisions or bail us out. No GPS to guide the way. We will have to be our own captain.

So (cue the music) may our decisions be wise.

May our futures fulfill our dreams.

May the courses we set for the strange new worlds that lie ahead be filled with surprise and mystery, but may we arrive safe and sound with our humanity intact.

And above all . . .

. . . may every one of us boldly go.

SUGGESTED READING AND SURFING

The books and Web sites listed here are not exhaustive and they certainly aren't meant to be a formal bibliography—just some suggestions that you may find interesting. If these lead you to other ideas, insights, and stories that you are compelled to explore further, our condolences. You have contracted the Wonderment Disease! Enjoy.

—W.S. and C.W.

Books

2001: A Space Odyssey, Arthur C. Clarke, New American Library.

A Brief History of Time, Stephen Hawking, Bantam Books.

The Age of Spiritual Machines, Ray Kurzweil, Viking.

Alan Turing: The Enigma, Andrew Hodges, Simon & Schuster.

The Anthropic Cosmological Principle, John D. Barrow and Frank Tipler, Oxford University Press.

Beyond Humanity: CyberEvolution and Future Minds, Gregory Paul and Earl Cox, Charles River Media.

The Creators, Daniel J. Boorstin, Random House/Vintage.

Disturbing the Universe, Freeman Dyson, Basic Books.

Engines of Creation, K. Eric Drexler, Doubleday.

The Fantasies of Robert Heinlein, Robert Heinlein, Tor.

Flatland, A Romance of Many Dimensions, Edwin Abbott, HarperCollins.

I, Robot, Isaac Asimov, Doubleday.

Infinite in All Directions, Freeman Dyson, HarperCollins.

Inside Star Trek, Herbert F. Solow and Robert H. Justman, Pocket Books.

The Lives of a Cell, Lewis Thomas,Viking Press.

The Making of Star Trek, Gene Roddenberry and Stephen E. Whitfield, Ballantine Books.

The Medusa and the Snail, Lewis Thomas,Viking Press.

Mind Children, Hans Moravec, Harvard University Press.

Mirror Worlds, David Gelernter, Oxford University Press.

Nanosystems, K. Eric Drexler, John Wiley and Sons.

Powers of Ten, Philip Morrison and Phylis Morrison, *Scientific American*.

Profiles of the Future, Arthur C. Clarke, HarperCollins.

Robot: Mere Machines to Transcendent Mind, Hans Moravec, Oxford University Press.

The Society of Mind, Marvin Minsky, Simon & Schuster.

Space Age, William J. (Chip) Walter, Random House.

Star Trek Chronology, Michael Okuda and Denise Okuda, Pocket Books.

Star Trek: Deep Space Nine Technical Manual, Herman Zimmerman, Rick Sternbach, Doug Drexler, Pocket Books.

The Star Trek Encyclopedia, Michael Okuda and Denise Okuda, Pocket Books.

Star Trek Science Logs, Andre Bormanis, Pocket Books.

Star Trek: The Next Generation Technical Manual, Rick Sternbach and Michael Okuda, Pocket Books.

Time Travel, Paul J. Nahin, Writer's Digest Books.

Unbounding the Future, K. Eric Drexler and Chris Peters with Gayle Pergamit, William Morrow and Co.

Voyage of the Space Beagle, A. E. Van Vogt, Simon & Schuster.

The Way Life Works, Mahlon Hoagland and Bert Dodson, Times Books.

Web Sites:

www.media.mit.edu
The MIT Media Laboratory's primary site. Nice overview of history, people, and projects.

www.parc.xerox.com/istl/projects/dlib
For insight into digital paper and other related innovations at Xerox PARC.

www.parc.xerox.com
PARC's primary site. Enter here for a look at everything they're doing.

www.foresight.org/homepage.html
The Foresight Institute's home page. This is the most authoritative site on nanotechnology, particularly molecular assembly, out there.

www.foresight.org/NanoRev/index.html
Excellent primer on nanotechnology and additional sources of information for both the general and technical reader.

spaceflight.nasa.gov/index.html
NASA site that focuses on the issues of human space flight and the technologies that make it possible.

www.cmu.edu
Primary Carnegie Mellon Web site. Overview of the university and all that it has to offer.

www.simonsays.com/st
Even tells you how to go about writing and submitting a *Star Trek* novel for publication.

slashdot.org
Self-proclaimed "News for Nerds." Offbeat view of the scientific world and what's going on there.

www.extropy.com
Primary site for the Extropy Institute, the International Transhumanist Organization. Fascinating and thorough look at how technology can be used to extend a healthy life, augment intelligence, improve your personal life, and optimize social systems. Not your average science site.

www.sciam.com
Primary site for *Scientific American* magazine. Excellent way to stay abreast of the most cutting edge science.

www.secularhumanism.org/library/fi/clarke_19_2.html
An illuminating chat with Arthur C. Clarke, one of the greats of science fiction literature and a top-notch futurist who has been right more often than most.

online.itp.ucsb.edu/online/plecture/thorne
Very cool site with a complete lecture (and drawings) presented by Kip Thorne on space/time warps.

www.etc.cmu.edu
The primary site for Carnegie Mellon's Entertainment Technology Center. Check it out.

www.transhumanist.com
Online journal with articles by smart people about the evolution of humans and technology. You may not agree with it, but it's thought provoking.

www.world.honda.com/robot
Check out this site to learn everything you'd like about Honda's Humanoid Robot project. (You can even rent one.)

www.arc.nasa.gov
Primary Web site for NASA's Ames Research Center.

www.hawking.org.uk/home/hindex.html
This is Stephen Hawking's personal Web site. Background, lectures, and mind-bending thoughts about the universe.

www.pbs.org/wnet/hawking/html/home.html
Web site created for a PBS series based on the theories and views of Stephen Hawking. Nice primer on the strange happenings in the universe and Hawking's view of it all.

www.treasure-troves.com/bios
Web site that features biographies of the greatest scientists who have ever lived, right up to the present.

www.nasa.gov
NASA's primary site. It's huge!

www.wearablegroup.org
Web site of the Wearable Computing Group at Carnegie Mellon featured in this book.

www.roddenberry.com
The official Web site for Gene Roddenberry.

www.aip.org/physnews/graphics/html/teleport.htm
American Institute of Physics site featuring an overview of the concept of quantum teleportation and entangled photons for those hoping that the transporter will soon be a reality. Fascinating.

www.foresight.org/nanomedicine
The Foresight Institute's site featuring a peek at nanomedicine.

www.lerc.nasa.gov/www/pao/warp.htm
NASA site where advanced propulsion expert Marc Millis takes you through an overview of what has been conceived, what scientists are thinking now, and what has to be overcome in the future to attain anything like light speed.

www.ornl.gov/TechResources/human_genome
Main information site for the Human Genome Project. Great place to start for an overview.

www.zyvex.com/nanotech/convergent.html
Ralph Merkle's complete explanation of how to build a molecular machine that can build other molecular machines.

www.nytimes.com/partners/microsites/chess/archive8.html
An article in *The New York Times* about Deep Blue, the machine that beat Gary Kasparov, that explores whether it's intelligent or not.

www.foresight.org/eoc/eoc_chapter_14.html
Online version of chapter 14 of Eric Drexler's groundbreaking book on nanotechnology, *Engines of Creation*.

www.merkle.com
Ralph Merkle's personal Web site and all interesting ideas thereto appended.

www.williamshatner.com
The author's personal Web site.

www.KurzweilAI.net/index.html?flash=1
Ray Kurzweil recently created this site for those who want to explore the mysteries and possibilities of artificial intelligence.

www.moma.org/exhibitions/un-privatehouse/contents.html
Site illustrating the Un-private House exhibition at the Museum of Modern Art discussed in this book. ("Things That Think.")

ACKNOWLEDGMENTS

When it comes to writing books, there is one thing that the Borg and we humans have in common: Can't do it alone. Although galactic domination wasn't my goal, I certainly couldn't have completed this book without the expertise, insight, time, and patience of legions of humanoids . . . well, at least scores of humanoids.

Our list of thank-yous is, therefore, voluminous, but heartfelt. . . .

First, my deepest gratitude to the people who helped us fathom and recall the rich, inventive, and mysterious technological world of *Star Trek* itself. Without that, this book would have made no sense. Nowhere has an imaginary future been so fully and marvelously developed . . . which may explain why much of it is now coming true.

One of the people who helped imagine it was Dorothy (D. C.) Fontana who was *Star Trek*-ing as Gene Roddenberry's assistant and then as one of the series' writers. Later when *Star Trek: The Next Generation* came into the world, she joined Gene as associate producer to add more insight and creativity. D. C. patiently sat with me and plumbed her impressive memory to shed light on what it was like thirty-five years ago as Gene and the staff shaped and wrestled with conceiving futuristic technologies like warp drive, the communicator, and the transporter. Her light, as usual, was bright and revealing.

Bob Justman did the same. It's arguable that *Star Trek* would never have made it on the air in the first place if it hadn't been for Bob, who as associate producer and later as co-producer of the series, wrangled *Star Trek*'s crew and aliens and staff day in and day out to make sure each episode was completed come hell or alien invasion. Nearly two decades later, like Dorothy, he helped get *The Next Generation* launched as that series's supervising producer. He applied his trademark thoroughness and patience to the task of rummaging through the emporium of his memory to recall the origins of ideas from the tricorder and the sickbay to the holodeck and warp drive. As always, Bob's insights were both delightful and detailed.

I especially appreciate the time that Stephen Poe gave me. Stephen (under the pseudonym Stephen Whitfield) cowrote *The Making of Star Trek* in 1968 with Gene Roddenberry. Even though he was suffering from leukemia when we spoke, he was generous and kind and a deep pool of information. The insights he provided as well as those in his book were extremely helpful. Unfortunately only a few months after we first talked, Stephen died. Yet he was kind and generous with his precious time to the end.

Special thanks to Michael Okuda too. No wonder he authored *The Star Trek Encyclopedia* (with his wife Denise). His knowledge of all things *Star Trek* truly is remarkable whether it was technological, artistic, or historical. More than once he tapped that enormous database between his ears to provide facts or leads that helped us track down the origins and inner workings of Data, the holodeck, dilithium crystals, and any number of other *Star Trek*-nologies.

Comprehending the imaginary technologies that *Star Trek* has envisioned was difficult enough for the likes of me, but the premise of this book could never have been fulfilled if it weren't for the scores of scientists who tolerated my questions about the real-life work they do which is today transforming *Star Trek*'s science fiction into science fact.

Special thanks go out to Kip Thorne, Hans Moravec, Elbert Branscomb, and Trevor Hawkins; Eric Drexler, Ralph Merkle, Ray Kurzweil, Lawrence Krauss, and Neil Gershenfeld. Every one of these scientists are leading thinkers in their fields and luminaries within the scientific community. Yet they and their colleagues took the time to talk with the former captain of a starship that never existed and to answer impertinent and naïve questions about particle physics, time travel, robotics, and the mysterious workings of molecules, cells, and computer chips. I never realized until this book exactly how fun it could be to hang out in the company of scientific genius.

Thanks to Kip for his mind-boggling insights into time travel so patiently and clearly passed along. Hans for the wild and thought-provoking theories and predictions about robotics and artificial intelligence. Elbert for his enthusiasm and talent for story-telling (which made the mysteries of the human genome so much less mysterious for me). Trevor for his deep pools of knowledge about the workings of DNA and the patience with which he explained them. Eric, not only for helping me understand the astounding world he envisions when nanotechnology revolutionizes the future, but for digressing so tolerantly to try to explain particle physics to me over lunch one afternoon. Ralph for his ferocious intellect, delicious sense of humor, and impressive talent for metaphor. Ray for his ability to make the inscrutable world of artificial intelligence, well, scrutable, yet endlessly fascinating. Lawrence for his rigorous, unique, and entertaining views on the science of teleportation, and Neil for his enviable flair for illustrating complicated ideas in such entertaining ways.

Not only did they all take the time to talk, but they willingly plowed through drafts of the book to make certain the considerable science it tackles (all of it well beyond me) was absolutely accurate.

I am equally thankful to Andy Berlin, Marc Millis, Chuck Jorgensen, Peter Norvig, Roy Want, Randy Pausch, Josh Hall, Dan Siewiorek, Francine Gemperle, Sebastian Thrun, Nick Roy, Mike Montemero, and Greg Armstrong. From smart matter to wearable computers, from robotics to virtual reality, their work and the work of their colleagues took this book into places I certainly could never imagine. Thanks for exposing me to your fascinating minds. I hope there won't be any radioactive fallout!

It took a lot of work to track down these stories and more work to arrange the meetings and organize and follow up on the information they produced. None of that could have happened without the help of some very kind and capable people. In particular, John Bluck at NASA Ames, Anne Watzman at Carnegie Mellon University, Deborah Cohen at the MIT Media Lab, and Lois Wong at Xerox PARC. Lois, in particular, not only helped us with many logistics, but also provided thoughtful comments on early drafts of the manuscript.

Both of my assistants during this project, first Stephanie Riggs and then Robin Guido were tireless in making phone calls, arranging travel and accommodations, putting up with stupid questions about where this was or how we could do that. Robin was relentless in tracking down and clearing rights to photos and quotes for inclusion in the book. They both put up with, and accomplished, a lot, despite my infernal interference.

Jane Singer took on the frightening task of transcribing every one of the interviews conducted for this book. This was not only insanely difficult because the discussions she transcribed were about some pretty esoteric science, but because, more often than not, three, four, even five people were involved in the discussions! Imagine listening to *that* on tape and trying to keep it all straight! Yet Jane did, *and* she often supplemented her work with invaluable insights that connected information in the interviews with historical *Star Trek* tidbits that had entirely escaped the notice of the authors.

Special personal thanks are in order to our agents at the Fifi Oscard Agency in New York, Carmen La Via and Peter Sawyer, who orchestrated the cosmic forces that made this book possible in the first place. They pushed, pulled, and massaged both people and proposals to pull the whole deal together. With their help, we skirted more than one black hole.

Many thanks to Mary Murrin Smith as well who read the book in various stages and offered never-ending encouragement and insight. When you sometimes feel you are writing nothing but gibberish, encouragement is very much appreciated.

Of course I want to thank my co-author, Chip Walter. Sometimes our journeys resembled a Crosby and Hope road show more than a Starfleet mission. As a result we had plenty of laughs from coast to coast, not to mention more than a few mind-bending conversations about what the future might hold. I went through some personally very difficult times during work on this book, and having Chip as a willing ear and first-class collaborator made the days a little less dark. Chip's unusually broad scientific knowledge and his insights into how the work of various scientists could be linked to *Star Trek* made writing this book far easier than it might have been. Never was mind-melding so much fun.